Information Theory in Analytical Chemistry

CHEMICAL ANALYSIS

A SERIES OF MONOGRAPHS ON
ANALYTICAL CHEMISTRY AND ITS APPLICATIONS

Editor
J. D. WINEFORDNER

VOLUME 128

A WILEY-INTERSCIENCE PUBLICATION

JOHN WILEY & SONS, INC.

New York / Chichester / Brisbane / Toronto / Singapore

Information Theory
in Analytical Chemistry

KAREL ECKSCHLAGER

Charles University of Prague

KLAUS DANZER

Friedrich Schiller University of Jena

A WILEY-INTERSCIENCE PUBLICATION

JOHN WILEY & SONS, INC.

New York / Chichester / Brisbane / Toronto / Singapore

543
E19 i

PREFACE

The development of analytics as a self-contained scientific discipline has occurred in two directions. On the one hand, it seeks various analytically relevant phenomena and proliferates the principles of analytical methods through laboratory instrumentation and computer techniques. On the other, it further elaborates the theory and derives practical consequences from it. Information theory has had an influence in both directions since the early 1970s, in particular in the development and application of analytical techniques that serve to assess and optimize general procedures and in fine-tuning the theoretical fundamentals of analytics. Recently information theory has also begun to be applied to the assessment of chemometric methods of multivariate data analysis. Given that an analytical procedure can comprise several operations whose order and interrelations affect the information gain of the analytical results, a systems approach to addressing the problems of contemporary analytics may be a necessary consideration as well.

This book is a natural sequel to Volume 53 in this monograph series on analytical chemistry and its application—namely the monograph *Information Theory as Applied to Chemical Analysis* by K. Eckschlager and V. Štěpánek (1979). In the present book we aim at more practical applications of information theory in analytical chemistry. We also center our discussion on the systems aspect of analysis—identification, qualitative, quantitative, surface, and structural analyses, all of which are important in multicomponent analysis. A separate chapter is devoted to the information contribution of multivariate data analysis, including cluster analysis, pattern recognition, and factorial analysis. The application of information theory to systems of creation and to the processing of chemical information relating to composition also enables us to demonstrate that multicomponent analysis or multivariate data analysis is more than a mere simultaneous multiple analysis of single components or

v

processing of isolated results. Significant interferences are possible, for instance, not only by the limited selectivity of the analytical procedure or properties of the instrument but also by the different relevant pieces of information obtained simultaneously. One can further show—often quantitatively—the effect of the individual unfavorable factors on the information gain of the analytical results. Of particular interest in this respect is the use of information theory in the detection of trace concentrations approaching the limit of determination. All these are actual problems of modern practical analytics that can be reasonably solved by a combination of information theory and the systems approach.

We would like to close this prefatory statement by thanking the editor of this series, Professor J. D. Winefordner of the University of Florida in Gainesville, who invited and urged us to write this book. We would also like to extend our thanks to our colleagues for their thoughtful comments on some of the ideas and research subjects; we thank especially Prof. K. Doerffel, Leipzig, Prof. G. Ehrlich, Dresden, and Dr. V. Štěpánek, Prague. Finally, we want to express our appreciation to the editors of this book for their assistance, especially in improving the English.

K. ECKSCHLAGER
Prague, Czech Republic

K. DANZER
Jena, Federal Republic of Germany

CONTENTS

CHEMICAL ANALYSIS

A SERIES OF MONOGRAPHS ON
ANALYTICAL CHEMISTRY AND ITS APPLICATIONS

J. D. Winefordner, *Series Editor*

Information Theory in Analytical Chemistry

CHAPTER

1

INTRODUCTION

The need for ever more information on chemical composition, its change over time, and spatial distribution has been growing steadily over the past decades. The demands put on the speed and efficiency of acquiring that information have also increased, as have those pertaining to the reliability and selectivity of analytical procedures. In particular, there is an urgent need to quantitate lower and lower amounts of analytes. All these requirements can be at least partly satisfied by diverse analytical methods and advanced analytical instrumentation and computer techniques. To meet these needs, analytical chemistry, which was formerly considered an integral part of experimental chemistry, has evolved into a methodologically differentiated, exact interdisciplinary scientific discipline that has benefited from the theoretical findings of many other sciences.

Mathematically defined notations that describe the theoretical foundations of various exact sciences are an important feature of analytical chemistry. Since the early 1970s concepts and definitions taken over from information theory and system theory have been applied to the analytical process (Malissa 1972). The concepts and notations of information theory play an integrating role in the universal methodologically differentiated analytics, independent of experimental chemical methods (Doerffel 1988). In addition information theory, along with the system approach, has demonstrated that the information contribution of an analytical result is largely determined by the purpose and logic behind analytical operations, an optimization of partial operations of the system, and the technical standard of the instrumentation.

Today our knowledge about the chemical and physical nature of partial processes and error factors is rather profound (Eckschlager 1969), so we no longer should consider the analytical system and its subsystems as a black box. Coming from the information-theoretic and system-theoretic points of view, the advances in chemical analytics can form the basis for the development of general rules of good laboratory practice (GLP), which may prove to be of use in the theory and practice of quality assurance (QA) as well. The information-theoretic basis of analytical chemistry along with the physicochemical principles of various analytical methods has already enabled an optimum chemometric strategy for solving specific problems (Doerffel, Eckschlager and Henrion 1990).

A more detailed discussion of the information-theoretic background of analytical chemistry can be found in Eckschlager and Stepánek (1979, 1985) and in reviews by Eckschlager and Stepánek (1982), Danzer, Eckschlager, and Wienke (1987), Danzer, Eckschlager, and Matherny (1989), Eckschlager, Stepánek, and Danzer (1990), Danzer, Schubert, and Liebich (1991), and Eckschlager (1991). A modern analytics textbook founded on this information-theoretic and system-theoretic foundation has been written by Danzer et al. (1987).

REFERENCES

Danzer, K., Eckschlager, K., and Matherny, M. 1989. *Fresenius Z. Anal. Chem.* **334**, 1.

Danzer, K., Eckschlager, K., and Wienke, D. 1987. *Fresenius Z. Anal. Chem.* **327**, 312.

Danzer, K., Than, E., Molch, D., and Küchler, L. 1987. *Analytik— Systematischer Überblick*. 2d ed Akademische Verlagsgesellschaft Geest & Portig, Leipzig.

Doerffel, K. 1988. *Fresenius Z. Anal. Chem.* **330**, 24.

Doerffel, K., Eckschlager, K., and Henrion, G. 1990. *Chemometrische Strategien in der Analytik*. Deutscher Verlag für Grundstoffindustrie, Leipzig.

Eckschlager, K. 1969. *Errors, Measurement and Results in Chemical Analysis*. Van Nostrand, London.

Eckschlager, K. 1991. *Collect. Czech. Chem. Commun.* **56**, 506.

Eckschlager, K., and Stepánek, V. 1979. *Information Theory as Applied to Chemical Analysis*. Wiley, New York.

Eckschlager, K., and Stepánek, V. 1982. *Anal. Chem.* **54**, 1115A.

Eckschlager, K., and Stepánek, V. 1985. *Analytical Measurement and Information. Advances in the Information Theoretic Approach to Chemical Analyses.* Research Studies Press, Letchworth.

Eckschlager, K., Stepánek, V., and Danzer, K. 1990. *J. Chemometrics* **4**, 195.

Malissa, H. 1972. *Automation in und mit der analytischen Chemie.* Verlag der Wiener Medizinischen Akademie, Vienna.

CHAPTER

2

THE AIM OF ANALYTICAL CHEMISTRY

The object of analytical chemistry is to obtain qualitative and quantitative information about the chemical composition and structure of materials. Analytical chemistry investigates the type of components present in materials, their amount, and their structural relationships.

Analytical investigations are directed at solving general problems. For this purpose, there have been developed methods of analysis that include sampling, sample preparation, and evaluation so that statistical and data analyses can be used in interpreting of the results. In effect the chemical analysis—the process of gaining information about the chemical composition of a sample—occurs within a stochastic system, that is, the outcome of successively repeated data follows a probability distribution.

Any analytical system comprises a set of subsystems in which the entire procedure is broken up into smaller operations—sampling, sample handling and preparation, separation, measurement, calibration, signal processing, result calculation, and so on. The order, interrelations, and feedback between the subsystems are critical to the functioning of the entire analytical system. Any errors in these subsystems, will contribute to the uncertainty of the final result. Thus, for the function of the entire analytical system to be efficient, not only must the chosen analytical method be suitable but also the subsystems must be optimized. The analytical system that is not stable over time may—if uncontrolled—exhibit a decreasing trend in information gain.

Information about chemical compositions is not directly measurable. It is derived from measurements of a material's analytical properties; the measurements disclose the kind and amount of analytes present and their structural arrangement. In any analytical investigation it is important that three main components—the **problem**, the **sample**, and the **method** used (shown in Fig. 2.1),—

5

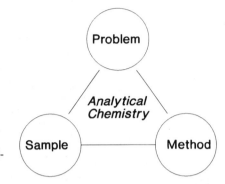

Figure 2.1. The three parts of an analytical investigation.

comply with each other in an optimum way. Each component of this "analytical trinity" (Betteridge 1976) is characterized by its special information-theoretic background (see Fig. 2.5).

2.1 THE ANALYTICAL PROCESS

Analytical work involves specific problems that can be scientific, engineering, or economical in nature. Even nontechnical questions may be of concern. To characterize an object exhaustively with regard to the given problem, a representative sample is taken. The main steps of analytical work, as well as a breakdown into subordinate steps, is shown in Fig. 2.2 (Danzer et al. 1987).

The analytical process outlined in the figure can be taken as the basis of all analytical investigations, whether they deal with identi-

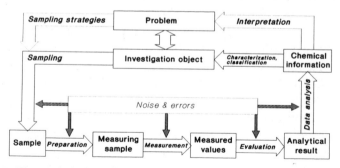

Figure 2.2. The analytical process.

fication or determination of components or overall structure. The analytical process attempts at obtaining information applicable to the solution of the task at hand. The analytical process starts with sampling and sample handling. By careful preparation and separation techniques, the sample is adapted to the demands of the measuring procedure.

A sample of matter is a potential carrier of information. It contains hidden information about its chemical composition and structure in a static form. According to Malissa (1984) a sample "... is not an ordinary piece or volume of matter, but all above it is a synholon, a syncretic body of matter and information." This information is extracted during an analytic process in which the sample is allowed to interact with other substances or is subjected to various forms of energy, and the signals that emerge are evaluated. The measurements established in the extraction of information are obtained from chemical reactions, electrochemical processes, and interactions of the sample with different energy forms, particularly radiation. Figure 2.3 gives a basic outline of the standard analytical measurements (Danzer 1992).

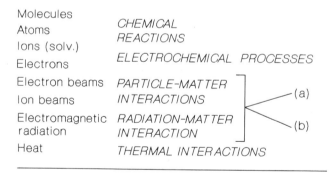

(a) inelastic interactions (with energy transfer):

SPECTROSCOPIC METHODS

(b) elastic interactions (without energy transfer):

DIFFRACTOMETRY, MICROSCOPY

Figure 2.3. Overview of the effects of different forms of matter and energy in interactions with samples.

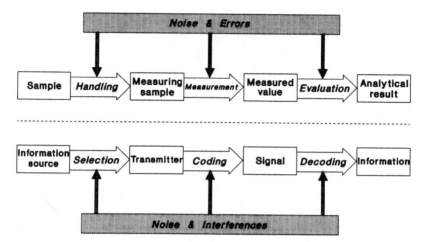

Figure 2.4. Comparison of the analytical process (*top*) with the general principles of processing information (*bottom*).

Often the analytical process is portrayed more or less as an information-processing chart, as shown in Fig. 2.4. The chemical analysis yields information, of which there are three kinds: **nominal, ordinal,** or **cardinal quantities**.

1. *Nominal quantities* (kind, identity, etc.) can be assigned names, symbols, formulas, or schemes, but they cannot be arranged in some order or represented by a numerical value. Nominal quantities are always discrete.
2. *Ordinal quantities* are discrete quantities that can be arranged in some order or classified into various levels to which numerical values may be assigned.
3. *Cardinal (quantitative) quantities* are expressed numerically in units. The numerical quantities can be fixed or random, continuous or discrete. By its nature, the value of any, even a fixed, cardinal quantity obtained by measurement or by chemical analysis is always a random quantity due to the error factor of measurement.

Qualitative analytical information can be expressed by nominal or ordinal quantities, whereas quantitative information is always represented by cardinal quantities. However, semiquantitative analysis, such as in optical emission spectrography (OES), quantitative information may also be expressed by ordinal quantities, such as $+++/++/+/0$.

The process occurring between the sampling and the evaluation of information about the chemical composition of the sample is usually both chemical and physical in nature. It involves work with laboratory equipment and instruments, standard samples, reference materials, and so on. Any measurement error carried over from the partial operations, which will affect the uncertainty of the final result, may now be located quantitatively and eliminated (Eckschlager 1969). To obtain undistorted signals, high technical standards must be maintained in the laboratory instrumentation and "quality standards" and reference materials must be followed. Moreover, state-of-the-art computer hardware and software, with a program in chemometrics, is critical for optimum decoding of the signals. Last, it is important to be able to assess qualitatively the effect of all factors on the results. A better picture of the approach to understanding the information about a chemical composition is developed next.

2.2   THE ANALYTICAL SYSTEM

The process of obtaining information about a chemical composition always proceeds within an analytical system. The analytical system here is a diffuse system (Nalimov 1971). A good model of a real diffuse system is the stochastic model in which repetition of the process for the same input always gives output values that obey some probabilistic distribution, generally the normal (Gaussian) distribution. Relationships within information systems are usually represented by **Venn diagrams**. A Venn graph for the analytical system we are considering is shown in Fig. 2.5 (Danzer 1979; Danzer, Eckschlager, and Wienke 1987). In the figure the diagram consists of three overlapping circles which represent the informa-

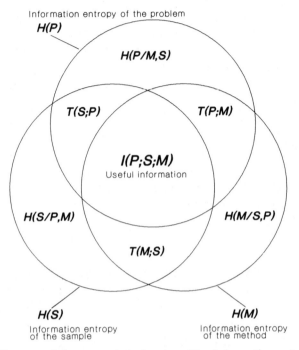

Figure 2.5. Venn graph for an analytical system. The information entropies $H(P)$, $H(M)$, and $H(S)$ refer to complete circles, and other symbols to corresponding sections.

tion entropies (H) of the problem $H(P)$, the sample $H(S)$, and the method $H(M)$.

At the intersection of all the information sections is *useful information*, $I(P; S; M)$, which is the most important because it includes the information about the problem, the method, and the sample. In other words, useful information is the only segment of the total entropy of the system that concerns the problem and the sample and is accessible by the method used. The transinformation between method and sample, $T(M; S)$ also provides real information about the sample, but what is there does not have any relevance to the problem. Here are, above all, the results of multicomponent analysis as, for example, in optical emission spectrography where the information would contain all of the elements in the sample from which only some may be of interest for a certain problem. Likewise the transinformation between sample and problem $T(S; P)$ is that part of information that is needed for solving the problem but that the method cannot yield. The overlapping area of useful information may be enlarged and gain in this information obtained by optimum sample handling (preparation steps, enrichment, etc.) and by the adaption of methods to the requirements of samples (e.g., solid sample atomic absorption spectrometry).

Some of the transinformation and entropy shares are of lower interest for the analyst, such as the transinformation $T(P; M)$ between problem and method. Such information may be of interest for problem solving and may be received, in principle, by the method used, but it is not contained in the sample. The same holds for the so-called conditional information entropies $H(M/S, P)$ of the method but irrelevant to the sample and problem, $H(S/P, M)$ of the sample but irrelevant to the problem and method, and $H(P/M, S)$ of the problem but irrelevant to the method and sample. The analytical system must be optimized to maximize useful information $I(P; S; M)$ and to reduce irrelevant transinformation and entropy shares.

In the early years, when information theory was just starting to be applied in analytical chemistry, analogies were sought between the model of an analytical system and the information transmission system. The information properties of analytical results and methods were characterized in terms of measures employed in communication theory. The measures based on Shannon's entropy (Shannon 1948; Shannon and Weaver 1949) were then well suited for use in qualitative and identification analyses (Cleij and Dijkstra 1979). However, there remained some quantitative results, particularly concerning multicomponent analysis, whose properties could not be characterized satisfactorily by information theory.

Work progressed until a model was developed that could describe the system in which the process of acquiring information on the chemical composition occurs (Eckschlager and Stepánek, 1985), beginning with the concept of the divergence measure (Eckschlager 1975) and later extending to the assessment of information properties of quantitative results (Eckschlager and Stepánek, 1985). That model will now be briefly described.

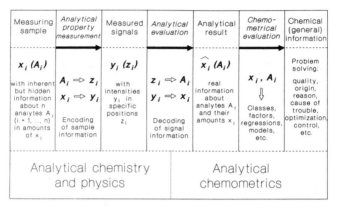

Figure 2.6. Process of encoding information from a sample into signals, decoding signals into analytical information, and extracting chemical information.

In some interpretations, the analytical system is construed as a system composed of a number of subsystems in which the partial operations of the analytical procedure take place. Between the subsystems there exist different, but exactly determined, relationships and feedbacks. The order of the operations is important to the function of the entire analytical system. Unlike the earlier analytical method, all the operations occurring in the subsystems must be evaluated and optimized (Doerffel, Eckschlager, and Henrion 1990).

Since information about chemical compositions cannot directly be measured, some other property that is closely related to the kind and amount of analytes is measured. This property is sometimes referred to as the *analytical property*. Chemical, physicochemical, and physical analytical properties are most frequently used, but biological properties are occasionally employed.

Thus the analytical system can be divided into three parts as shown in Fig. 2.6: first, property measurements are encoded into signals, second, the signals are decoded into the quantitative results, and third, chemometric methods are used to extract chemical information from the results which are interpreted to answer the general problem at hand.

The subsystems between the input and measurement are necessary for creating the signal, and subsystems between each measuring instrument's output and the output of the entire system provide for the decoding, signal processing, and transformation into the needed information. Stated in other words, the way the analytical signal is formed is the subject of analytical chemistry and physics, whereas the signal processing, decoding and interpretation of the analytical result is the task of analytical chemometrics. Within the analytical system there should continually be feedback between the second (evaluation) and the third (interpretation) parts. These control the function of the first (signal creation) part.

Information content (as information-theoretic quantity) refers to the first step of the analytical system and characterizes, therefore, the measured values. On the other hand, the terms *information gain* and *information contribution*, respectively, in the case of single-component analysis (or identification) and *total amount of information* in the case of multicomponent analysis characterize the entire analytical system. These quantities will be discussed later, but details can be found in publications by Eckschlager and Stepánek (1975, 1985), Cleij and Dijkstra (1979), Frank, Veress, and Pungor (1982), and Eckschlager, Stepánek, and Danzer (1990).

In most cases chemical information is the principal aim of the analytical investigation. Then the chemometrical evaluation provides the most useful information; this is the transinformation between problem, sample, and method $I(P, S, M)$, as shown in Fig. 2.5. Of course this is higher-level information. Analytical results serve to provide a classification, a statement, a decision on quality, and so on. Information obtained by chemometrical evaluation will be discussed in Chapter 12.

2.3 THE ANALYTICAL SIGNAL

The analytical signal, which carries information about a chemical composition, is always connected with a change in the chemical or physical state of a material. The signal can be transformed into analytical information if it fulfills **syntactic, semantic, pragmatic, and denotation functions**.

1. The *syntactic function* describes the relationship between equivalent signals—specifically concerning their origins or sequences and their transformations (e.g., laws of analytical processes originating in the signal).
2. The *semantic function* describes the meaning and content of a signal and thus its unambiguous connection with the object characterized—namely the problem of coding and decoding, assignment rules (e.g., spectral collections and tables), calibration and analytical functions.
3. The *pragmatic function* determines the relationship between a signal and the person who produces and, above all, receives it. It concerns the different meaning and the relative value of each analytical signal (e.g., measured value or the content of chemical reports) for different receivers.[1]
4. The *denotation (sigmatic) function* describes the relationship between a signal and its information content. Where a signal has an unambiguous meaning (which is a rule in chemical analysis) this function is considered a *semantic function*.

The syntactic, semantic, and pragmatic functions must be harmonized for the obtained information to be unambiguous and not allow misinterpretation. But only the syntactic and, to a very limited extent, semantic functions can be expressed mathematically. Careful attention must be paid to interpreting the semantic and pragmatic functions by including research objectives and a clear statement of the problem to be solved by the analytical process. Some partial views on this topic were published by Malissa as early as 1971.

Signals can appear as one-dimensional (i.e., possess intensity y only) or two-dimensional, (i.e., have intensity y in positions z, as in

---

[1]Each one of us knows what green and red means on traffic lights, but what might a flashing blue or violet light mean? We have no information here because there is no pragmatic function. How very different is the analytical result of 0.55 ± 0.04 wt-ppm Cu for a chemist faced with carrying out an analysis, on the one hand, and for his or her laboratory's janitor, on the other hand.

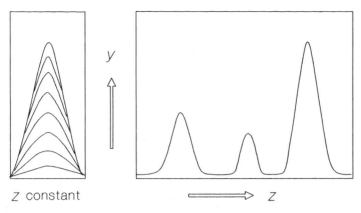

Figure 2.7. One-dimensional signal and two-dimensional functions.

$y(z)$). A signal is said to be one-dimensional if there exists no variability in z (i.e., in the case of chemical assay or determination) or if the intensity y is *measured* by a fixed value of z (i.e., in the case of colorimetric or photometric measurements). One- and two-dimensional signal functions are represented in Fig. 2.7.

Suppose that the signal-carrying information on a chemical composition has a position z, which contains information on the qualitative composition (i.e. on the identity of the ith analyte A_i, $i = 1, \ldots, n$), and can take continuous or discrete values (z_j, $j = 1, \ldots, k$; as a rule $k \geq n$). Further suppose that the signal has some intensity y or some intensity in some position $y(z_j)$ containing information on the amount or content x_i of ith analyte A_i in the sample.

A signal's position can be either a fixed quantity (a spectral line's wavelength) or depend on the conditions in which the analysis is carried out (retention quantities in chromatography). The measured value of the signal's position is always a continuous random quantity with a probability density $p(z)$, and the signal's intensity measured value is always a random quantity, which can be either discrete with a frequency function $f(y)$ or continuous with probability density $p(y)$.

Examples of variables z and y for various analytical methods are shown in Table 2.1. A discussion of signal shapes can be found in the work by Eckschlager and Stepánek (1979); signal processing is

Table 2.1. Signals of Some Analytical Methods

Signal	Method	Signal Position z	Signal Intensity y
z indirectly relevant	Gravimetry	Precipitating reagent, reaction conditions	Mass of the precipitate
	Titration	Titrant, reaction conditions	Volume of the volumetric solution
z = constant	Photometry	Absorption maximum	Absorbance
	Atomic absorption spectrometry	Wavelength of the resonance line	Absorbance
$y = f(z)$	DC polarography	Half-wave potential	Diffusion current
	Optical emission spectrography	Wavelength of the spectral line	Intensity of the spectral line
	Mass spectrometry	Mass number (m/e)	Peak height
	GC, HPLC	Retention time (volume)	Peak area

the subject of Danzer, Hopfe, and Marx (1982) and Doerffel, Eckschlager, and Henrion (1990). We will discuss the properties of signals in a later chapter on multicomponent analysis.

2.4   ANALYTICAL RESULT AND ANALYTICAL INFORMATION

Chemical analysis takes place in a system whose input is a material sample and output is information. The sample, as a representative of the material being analyzed is a carrier of latent information about its chemical composition. The information is extracted by creating a signal by reaction with a reagent or by interacting it with energy, followed by signal decoding. Since a *single sample*[1] cannot provide information on the spatial distribution or time changes of the chemical composition over an entire object, one has to distinguish between the following.

1. *Analytical result.* The information on the chemical composition of just the sample (e.g., an *average content*).
2. *Analytical information.* The way that the composition will change over time and its *spatial arrangement* within the material.

The analytical result always provides an answer to the question, "**What?**" or "**How much?**" (Eckschlager 1991). Analytical information must consider additional questions about the nature of the whole material system. Analytical information, which is obtained by chemometric processing, involves giving a logical interpretation after analyzing a series of samples taken from an appropriate region of interest. Computer techniques and software are frequently employed (Beebe and Kowalski 1987).

Suppose that we consider an analytical result that is a function of the quantities z and y, where z is the characterizing **kind** of analyte (kind-specific quantity), and y is the characterizing **amount** of analyte. To obtain analytical information in this case, we add the

---

[1]In its general meaning a sample is a segment of a large object under investigation, or on a small scale, it can be taken from a spot by means of a microprobe analyzer.

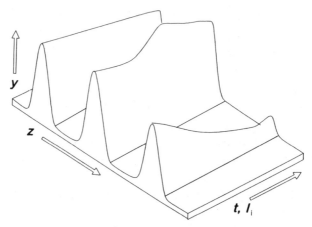

Figure 2.8. Analytical information $y = f(z)$ for time t or spatial coordinate l_i.

spatial coordinates l_1, l_2, and l_3, and we may even want to consider time t or some time-dependent parameter. An overview of different kinds of analytical information and corresponding functions can be found in Danzer, Eckschlager, and Wienke (1987).

Any analytical information has a factual aspect (assumptions) and a quantitative aspect (number of details, its size). Information theory deals only with the quantitative aspect of analytical information. We cannot assess quantitatively its factual aspect or its truthfulness, but we can take into account the relevancy of the experimentally obtained information when solving a specific problem (Eckschlager and Stepánek 1985, 1987). To some extent we can also assess the plausibility of the result obtained (i.e., its agreement with a priori assumptions) and prove an assumption based on some hypothesis (Eckschlager, Stepánek, and Danzer 1990). This approach will be illustrated later.

REFERENCES

Beebe, K. R., and Kowalski, B. R. 1987. *Anal. Chem.* **59**, 1007*A*.

Betteridge, D. 1976. *Anal. Chem.* **48**, 1034*A*.

Cleij, P., and Dijkstra, A. 1979. *Fresenius Z. Anal. Chem.* **298**, 97.

Danzer, K. 1979. *Zur Anwendung informationstheoretischer Grundlagen in der Analytik*. Habilitation thesis, Technical University Chemnitz.

Danzer, K. 1992. *Fresenius J. Anal. Chem.* **343**, 827.

Danzer, K., Eckschlager, K., and Matherny, M. 1989. *Fresenius Z. Anal. Chem.* **334**, 1.

Danzer, K., Hopfe, V., and Marx, G. 1982. *Z. Chem.* **22**, 332.

Danzer, K., Than, E., Molch, D., Küchler, L., and König, H. 1987. *Analytik. Systematischer Überblick*. 2d ed. Akademische Verlagsgesellschaft Geest & Portig, Leipzig.

Doerffel, K., Eckschlager, K., and Henrion, G. 1990. *Chemometrische Strategien in der Analytik*. Deutscher Verlag für Grundstoffindustrie, Leipzig.

Eckschlager, K. 1979. *Errors, Measurements and Results in Chemical Analysis*. Van Nostrand, London.

Eckschlager, K. 1975. *Fresenius Z. Anal. Chem.* **277**, 1.

Eckschlager, K. 1991. *Collect. Czech. Chem. Commun.* **56**, 506.

Eckschlager, K., and Stepánek, V. 1979. *Information Theory as Applied to Chemical Analysis*. Wiley, New York.

Eckschlager, K., and Stepánek, V. 1985. *Analytical Measurement and Information. Advances in the Information Theoretic Approach to Chemical Analysis*. Research Studies Press, Letchworth.

Eckschlager, K., and Stepánek, V. 1987. *Chemom. Intell. Lab. Syst.* **1**, 273.

Eckschlager, K., Stepánek, V., and Danzer, K. 1990. *J. Chemometrics*, **4**, 195.

Frank, I., Veress, G., and Pungor, E. 1982. *Hung. Sci. Instr.* **54**, 1.

Malissa, H. 1971. *Fresenius Z. Anal. Chem.* **256**, 7.

Malissa, H. 1984. *Fresenius Z. Anal. Chem.* **319**, 357.

Nalimov, V. V. 1971. *Teoriya Eksperimenta*. Nauka, Moscow.

Shannon, C. E. 1948. *Bell Syst. Techn. J.* **27**, 379, 623.

Shannon, C. E., and Weaver, W. 1949. *The Mathematical Theory of Communication*. University of Illinois Press, Urbana.

CHAPTER

3

BASIC CONCEPTS OF INFORMATION THEORY

The system in which chemical analysis takes place is a diffuse system in that the information obtained always has some uncertainty. This uncertainty may be due to imprecise measurements and errors in other quantitative results or to insufficient identification or resolvability in the qualitative interpretation. As pointed out by Gauss, the occurrence of measuring errors does not necessarily indicate a flaw, errors occur according to natural law.

Uncertainty is often inevitable, since random effects can occur during an experiment. The uncertainty in experimentally obtained information is therefore accounted for by probability theory. The cause of the uncertainty is seldom identified; it may involve a number of causes, some of which may never be known. So rather than look for causes, we consider the system as stochastic, whereby the output for one repeated input yields a set of data obeying some probability distribution. This distribution determines the **a posteriori uncertainty** that remains after the analysis. For us to be able to carry out analysis of the sample, we must have some preliminary information about its composition. Such preliminary information is said to have some **a priori uncertainty**. A priori uncertainty is higher than a posteriori uncertainty, and it can be characterized by means of the a priori probability distribution. Thus, both the a priori and the a posteriori uncertainties, are determined by their probability distributions.

The objective of chemical analysis is to obtain information about the chemical composition of a representative sample of the object in question, that is, to reduce the uncertainty about its chemical composition. By Bayesian reasoning (Vajda and Eckschlager 1980), the reduction of uncertainty can be expressed as the difference between the a priori uncertainty H_0 and the a posteriori uncertainty H:

$$I = H_0 - H \qquad (3.1)$$

The a posteriori uncertainty remaining after the analysis often consists of the dispersion caused by the variability of the analytical method, rather than the unremoved part of the initial uncertainty. The value of information in an analytical result (e.g., of a qualitative assay or a quantitative determination) depends on the way the a priori and a posteriori uncertainties are expressed in the situation.

If several components are determined simultaneously, the total amount of information M obtained from such multicomponent analysis is found as

$$M = \sum_{j=1}^{n} I_j \qquad (3.2)$$

where I_j is the information contribution of the assay or determination of analyte A_j ($j = 1, 2, \ldots, n$). Equation (3.2) assumes that the results are uncorrelated and of similar relevance. Basic terms and definitions were given by Malissa, Rendl, and Marr (1975).

3.1 PROBABILITY

The theory of probability applied to random experiments is based on the phenomenon of the stability of the relative frequency of the outcome of experiments. A random experiment provides, due to the effect of random circumstances, results that may be different but that follow a distribution whose shape may be known in advance. We cannot determine in advance which results we will obtain, since random experiment can be repeated as desired without the repetitions affecting each other. However, the results of a random experiment, even if repeated myriad times, are not chaotic; rather, there exists the phenomenon of statistical regularity: The relative frequency of each result of a random experiment in a certain arrangement approaches a constant which is referred to as the probability.

Mathematical statistics, in an inductive way, draws conclusions by which the findings obtained from a set of results of random experiments can be generalized. In the case of chemical analysis, the process of gaining information in a stochastic system can be regarded as a random experiment and the analytical result as a outcome of the random experiment. The fundamentals of probability theory and of the statistical inference method are outlined in a monograph by Eckschlager and Stepánek (1979 chs. 3 and 5). We will discuss here some of the basic concepts.

First, there exist two basic definitions of probability: *mathematical* and *statistical*.

3.1.1 Mathematical Definition

Problems in mathematical probability theory are formulated with respect to the set of all the possible results of a given random experiment. This set is referred to as the sample space S; its elements are the elementary events. Each subset A_S of the sample space is an event, the empty set 0 is an impossible event, and the entire sample space S is a certain event. The set function P defined for the class of all events A_S is the probability measure, and $P(A)$ is the probability of event A if it satisfies the following three axioms:

1. $P(A) \geq 0$ for all A's $\in A_S$.
2. $P(S) = 1$ (the probability of a certain event is unity).
3. $P(A \cup B) = P(A) + P(B)$ for disjoint A, B (i.e., for A $\cap$ B = $\varnothing$).

The probability space $\{S, A_S, P\}$ is the mathematical model of the random experiment.

This definition assumes that the event field contains a finite number of random events A_S. The so-called classical probability field, whose elementary events are equally probable, is a particular case. Then, the probability event A is

$$P(A) = \frac{n(A)}{N} \tag{3.3}$$

where $n(A)$ is the number of favorable cases and N is the number of all possible cases. The calculation of probability in a classical probability field is often based on combinatorial considerations; in many cases this is also feasible for the determination of the information content of results of quantitative experiments.

3.1.2   Statistical Definition

The statistical (von Misses') definition of probability requires neither a priori knowledge of the objective properties of the random phenomenon in question nor a finite number of elementary events. It is based on a multiple repetition of the random experiment. If the random experiment has been n-fold repeated and event A occurred $n(A)$-times, then

$$P(A) = \lim_{n \to \infty} \frac{n(A)}{n} \qquad (3.4)$$

That is, the relative frequency of event A converges to its probability for many repetitions of the random experiments.

Given the probability space $\{S, A_S, P\}$ where two events A, B obey $A \cap B \neq \varnothing$, $P(A) > 0$, $P(B) > 0$, then the conditional probability function

$$P(A|B) = \frac{P(A \cap B)}{P(A)}$$

$$P(B|A) = \frac{P(A \cap B)}{P(B)} \qquad (3.5)$$

attributes, for example, to event A the conditional probability $P(A|B)$ that event A will occur if event B has occurred. For $A \cap B = \varnothing$, we have $P(A|B) = P(B|A) = 0$.

In experiment theory cases are frequent where $A \subset B$, $A = B$, or $B \subset A$. Then the conditional probability is as follows:

$A \subset B$: $P(A|B) = P(A)/P(B)$, $P(B|A) = 1$.
$A = B$: $P(A|B) = P(B|A) = 1$.
$B \subset A$: $P(A|B) = 1$, $P(B|A) = P(B)/P(A)$.

Bayes's relation

$$P(A_1|B) = \frac{P(A_1)P(B|A_1)}{\sum_{i=1}^{n} P(A_i)P(B|A_i)} \tag{3.6}$$

enables the conditional probability $P(A|B)$ to be calculated from the $P(A)$ and $P(B|A)$ values. In cases where the cause and the consequence can be discriminated, Bayes's relation can be regarded as the theorem of probability of causes. Conditional probability is of significance in the qualitative experiment theory: It characterizes, for example, the probability $P(A_i|z_j)$ that sample contains analyte A_i if signal occurs at position z_j.

3.2 RANDOM QUANTITY

A quantity is referred to as random if, due to the effect of random phenomena, it takes different values with the same probability or with different probabilities. For instance, in throwing a die the probability of all the possible results $i = 1, 2, \ldots, 6$ is the same, namely $P(i) = 1/6$. In throwing two dice, the results—sum of spots of them—have different probabilities: For instance, the sum $i = 2$ can only be obtained from 1's on both dice, whereas the sum 7 can result as $1 + 6$, $2 + 5$, $3 + 4$; and since the number n of all possibilities in throwing two dice is $N = 6^2 = 36$, we have $P(2) = 1/36$ and $P(7) = 3/36 = 1/12$. Throwing dice is a random experiment, but the probability distribution for the various results depends on the arrangement of the experiment, for instance, on how many dice are thrown simultaneously.

The probability distribution for a random quantity is given by the describing function: Most frequently used are the cumulative distribution function and the probability function for discrete random quantities, and the probability density for continuous random quantities.

The distribution function of a random quantity ξ is the function $F(x) = P(\xi \geq x)$ having the following properties:

$$\lim_{x \to -\infty} F(x) = 0$$

$$\lim_{x \to +\infty} F(x) = 1 \tag{3.7}$$

The distribution function is nondescending and everywhere continuous from the right.

The distribution function obeys the following:

$$P(x_1 < \xi \le x_2) = F(x_2) - F(x_1)$$
$$P(\xi > x) = 1 - F(x)$$
$$P(\xi = x) = F(x) - F(x - 0)$$

If $F(x)$ is continuous in point x, then $P(\xi = x) = 0$; if $F(x)$ has a jump in point x, $P(\xi = x)$ represents the height of this jump.

A discrete (discontinuous) random quantity is a random quantity whose distribution function has a stepwise shape and the jumps in points x_i have heights $P(\xi = x_i)$. The distribution function is

$$F(x) = \sum_i P(\xi = x_i) \qquad (x_i \le x) \tag{3.8}$$

Function $f(x_i) = P(\xi = x_i)$ is referred to as the probability function of a discontinuous random quantity.

A continuous random quantity is a random quantity whose distribution function is continuous and has derivatives $p(x) = dF(x)/dx$ everywhere or nearly everywhere. The distribution function $F(x)$ is determined from the probability density $p(x)$ as

$$F(x) = \int_{-\infty}^{x} p(x)\, dx \tag{3.9}$$

For $p(x) \ge 0$ we have $\int_{-\infty}^{\infty} p(x)\, dx = 1$. Some continuous and discontinuous distribution and probability functions or probability densities are shown in Fig. 3.1.

Any probability distribution can be characterized by means of **moments**. Moments express certain properties of the distribution in a compressed form. The moment of the rth degree ($r > 0$, integer) is

$$m_r = \sum_i x_i^r f(x_i) \tag{3.10}$$

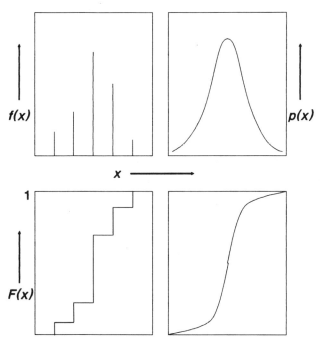

Figure 3.1. Discontinuous $f(x)$ and continuous $p(x)$ probability density functions (*top*) and distribution functions $F(x)$ (*bottom*)

for a discrete random quantity ξ and

$$m_r = \int x^r p(x)\, dx \qquad (3.11)$$

for a continuous random quantity ξ.

In addition to these general moments, the following *central moments* are also used:

$$m(\mu)_r = \sum_i (x_i - \mu)^r f(x_i) \qquad (3.12)$$

for a discrete random quantity ξ and

$$m(\mu)_r = \int (x_i - \mu)^r p(x)\, dx \qquad (3.13)$$

for a continuous random quantity ξ.

The most important are the first moment, $m_1 = E[\xi]$ (i.e., the expected value), which characterizes the position of the distribution on the scale of real numbers, and the second central moment, $m_2(\mu)_2 = V[\xi]$ (i.e., the dispersion variance of the probabilistic distribution), which characterizes the dispersion about the mean value. The root of the variance is the **standard deviation**.

We have for the expected value and for the variance

$$E[a] = a, \quad V[a] = 0 \,(a = \text{const.})$$
$$E[a \cdot \xi] = a \cdot E[\xi], \quad V[a \cdot \xi] = a^2 \cdot V[\xi]$$

3.3   TWO-DIMENSIONAL RANDOM QUANTITIES

Consider a pair of random quantities ξ and η. The joint distribution function

$$F(x, y) = P(\xi \le x, \eta \le y)$$

is the probability of simultaneous occurrence of both events for each pair x, y. If the joint distribution function of two random quantities is known, the individual distribution functions are

$$F(x) = F(x, \infty)$$
$$F(y) = F(\infty, y) \tag{3.14}$$

and they are referred to as the **marginal distribution functions**. The joint distribution function for continuous random quantities obey the relation

$$F(x, y) = \int_{-\infty}^{x} \int_{-\infty}^{y} p(x, y)\, dx\, dy \tag{3.15}$$

where $p(x, y)$ is the joint probability density. The random quantities ξ, η are stochastically independent if the following holds true for each pair (x, y):

$$F(x, y) = F(x) \cdot F(y)$$
$$p(x, y) = p(x) \cdot p(y)$$

For independent ξ and η we have

$$E[\xi \pm \eta] = E[\xi] \pm E[\eta]$$

$$V[\xi \pm \eta] = V[\xi] + V[\eta] \qquad (3.16)$$

The expected values are added/subtracted for addition/subtraction of the independent random variables, whereas the variances are added in either case.

The conditional probability density is

$$p(x|y) = \frac{p(x, y)}{p(y)} \qquad (3.17)$$

and the relation $p(x, y) = p(x|y) \cdot p(y)$, $p(y) > 0$, holds true.

The joint probability densities $p(x, y)$ are of importance in the determination of the information that the value of one random quantity (e.g., signal y) provides on the value of the other random quantity ξ (e.g., the analyte content).

3.4 SOME SPECIFIC PROBABILITY DISTRIBUTIONS

Many random experiments can be characterized by basically the same mathematical model if the ground rules of the experiment are reduced to common terms. Although the values of certain parameters vary from situation to situation, the basic structure of the model and the forms of the describing functions remain the same. Such models are important in probability and information theory applications, and becoming familiar with them is necessary.

The statement that a random variable has a certain probability distribution means that its describing function has a prescribed mathematical form. Knowledge of the distribution or its assumption lie in the basis of statistical inference. The origin of some probability distributions can be explained by a suitable physical experiment.

The **uniform**, or **rectangular**, **distribution** $U(x_1, x_2)$ is useful in applications because it implies equally likely outcomes or no prior choice. ξ is a continuous random variable having a uniform distribution on the interval $\langle x_1, x_2 \rangle$ if its probability density is given by

$$p(x) = \begin{cases} \dfrac{1}{x_2 - x_1} & \text{for } x_1 \leq x \leq x_2 \\ 0 & \text{otherwise} \end{cases} \tag{3.18}$$

The uniform distribution has the expected value

$$E[\xi] = \frac{x_1 + x_2}{2} \tag{3.19}$$

and the variance

$$V[\xi] = \frac{(x_2 - x_1)^2}{12} \tag{3.20}$$

Sketches of the probability density and the distribution function are shown in Fig. 3.2.

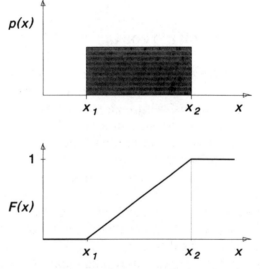

Figure 3.2. Probability density (*top*) and distribution function (*bottom*) of the uniform (rectangular) distribution

The **normal**, or **Gaussian**, distribution $N(\mu, \sigma^2)$ merits the first place among continuous distributions because it is the most important distribution in probability theory. It appears as the distribution of various sample statistics in statistical inference and is the limiting distribution of some other probability distributions. A typical example of the normal distribution is the distribution of random errors in measuring continuous quantities. Similarly the normal distribution arises when repeating chemical analysis of the same sample or when repeating a measurement under the same conditions. A random variable ξ has a normal, or Gaussian, distribution if its probability density has the form

$$p(x) = \frac{1}{\sigma\sqrt{(2\pi)}} \exp\left[-\frac{1}{2}\left(\frac{x-\mu}{\sigma}\right)^2\right] \qquad (3.21)$$

Thus it involves expectation μ and variance $\sigma^2 > 0$. The plot of this probability density is a bell-shaped curve (Fig. 3.3).

When $\mu = 0$ and $\sigma = 1$, we have the probability of the standardized normal random variable ζ; its distribution function (Fig. 3.3) is

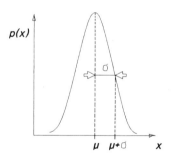

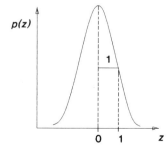

Figure 3.3. Gaussian distribution in its original and standardized form

specifically denoted $\phi(z)$:

$$\phi(z) = \frac{1}{\sqrt{(2\pi)}} \int_{-\infty}^{z} \exp\left(-\frac{z^2}{2}\right) dz \qquad (3.22)$$

A table of this function as well as of the probability density

$$\varphi(z) = \frac{1}{\sqrt{2\pi}} \exp\left(-\frac{z^2}{2}\right) \qquad (3.23)$$

can be found in any monograph on probability and statistics. Probability defined in terms of the distribution function $F(x)$ of any Gaussian random variable ξ or in its standardized form ζ according to the transformation

$$z = \frac{x - \mu}{\sigma} \qquad (3.24)$$

can be calculated from $\phi(z)$ as

$$F(x) = \phi(z) = F\left(\frac{x - \mu}{\sigma}\right) \qquad (3.25)$$

The probability density $\varphi(z)$ has its maximum at $z = 0$ and inflection points at ± 1.

Two other distributions that are important in the application of information theory to analytical chemistry are derived from the normal distribution. These are the **logarithmic-normal** (lognormal) and **truncated normal distributions**.

The logarithmic-normal distribution $LN(\mu, \sigma^2)$ is a distribution of a random variable ξ such that $\ln \xi$ has a normal distribution with parameters μ and σ^2. Its probability density is

$$p(x) = \begin{cases} 0 & x \leq 0 \\ \dfrac{1}{x\sigma\sqrt{2\pi}} \exp\left[-\dfrac{1}{2}\left(\dfrac{\ln x - \mu}{\sigma}\right)^2\right] & x > 0 \end{cases} \qquad (3.26)$$

The expected value $E[\xi] = \exp(\mu + \sigma^2/2)$ and the variance $V[\xi] = \exp(2\mu + \sigma^2)(\exp \sigma^2 - 1)$ are not mutually independent.

The shifted logarithmic-normal distribution $LN(\mu, \sigma^2; x_0)$, with its origin at $x_0 \geq 0$, is frequently better suited for practical uses. Its probability density is

$$p(x) = \begin{cases} 0 & x \leq x_0 \\ \dfrac{1}{(x - x_0)\sigma\sqrt{2\pi}} \exp\left[-\dfrac{1}{2}\left(\dfrac{\ln(x - x_0) - \mu}{\sigma}\right)^2\right] & x > x_0 \end{cases}$$

$$(3.27)$$

The expected value is $E[\xi] = x_0 + \exp[\mu + \sigma^2/2]$. Some other details concerning the shifted logarithmic-normal distribution can be found in the monograph by Eckschlager and Stepánek (1985).

The truncated normal distribution $TN(\mu, \sigma^2; x_0)$ is a normal distribution truncated in point x_0. Its probability density is

$$p(x) = \begin{cases} 0 & x \leq x_0 \\ \dfrac{1}{[1 - \phi(z)]\sigma\sqrt{2\pi}} \exp\left[-\dfrac{1}{2}\left(\dfrac{x - \mu}{\sigma}\right)^2\right] & x > x_0 \end{cases} \quad (3.28)$$

where $\phi(z)$ is the distribution function of the normal distribution for the random variable $z = (x_0 - \mu)/\sigma$ according to Eq. (3.24). The shape of the truncated normal distribution is shown in Fig. 3.4.

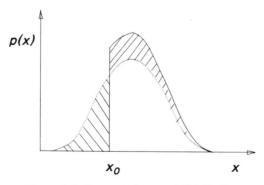

Figure 3.4. Truncated normal distribution

The expected value of the truncated normal distribution is $E[\xi]$ $= \mu + \sigma.\ \varphi(z)/[1 - \phi(z)]$, where $\varphi(z)$ is the probability density according to Eq. (3.23) for random variable z. Additional details can be found in the monograph by Eckschlager and Stepánek (1985) as noted above.

3.5   METHODS OF STATISTICAL INFERENCE

The basic objective of mathematical statistics is to present inference conclusions obtained by observing the set of values of one or several random variables. The set of results of all possible repetitions of a random experiment is called a **population**. For our needs in analytical chemistry problem solving, infinite and hypothetical populations are of most importance.

Clearly it is impossible to examine such populations in their entirety. Therefore we try to investigate them on the basis of their parts, that is, using **random samples**. For instance, if we determine the value of a random quantity ξ in a series of n independent repetitions of a random experiment, we obtain a random sample $(x_1, x_2, \ldots, x_n)$, where $x_i, i = 1, 2, \ldots, n$, are realizations of the random variable ξ with a distribution function $F(x)$.

A random sample must satisfy two basic conditions:

1. It must be representative, in that each element of the basic set must have the same chance to get into the sample.
2. The elements of the sample must be mutually independent.

While the moments μ and σ^2 of the basic set are fixed values, which for the distribution function $F(x)$ are given by Eqs. (3.10) through (3.13), the sample characteristics (sample "statistics") such as the sample mean or sample variance are random quantities that have probability distributions. It is clear that the more accurate the estimate of parameters of the basic set (population) is, wider is the extent of sample n.

The methods of statistical inference can be found in any mathematical statistics textbook, and the most important of them are briefly dealt with in the monograph by Eckschlager and Stepánek

(1979). What is important to the ensuing treatment is that one must discriminate between the values of probability distribution parameters such as $E[\xi] = \mu$ and $V[\xi] = \sigma^2$ and their sample estimates $\hat{\mu}$ and $\hat{\sigma}^2$.

Whereas information theory quantities are always defined for parameters of the a priori and a posteriori distributions used, only their estimates are available for practical calculations. Furthermore, although in determining the statistical contribution (see Section 3.8) we do not seek whether or not the bias is statistically significant, establishing the statistical significance of the difference between the true and found values is critical for the subsequent decision. Finally, establishing the tightness of correlation between two random variables η and ξ is important when assessing how reliable information on the other variable the one variable carries; this is called *mutual information*. Mutual information is a fundamental concept in analytical chemistry, since analytical information is usually sought by measuring quantities that relate closely to the analyte identity or concentration (see Sections 2.3 and 2.4).

3.6 INFORMATION UNCERTAINTY

The uncertainty of the result of a random experiment can be defined in terms of the probability of its occurrence in the discrete case, and in terms of its probability density in the continuous case. The uncertainty of a nominal, ordinal or cardinal quantity, which takes various possibilities, levels, or discrete values $i = 1, 2, \ldots, n$ with probabilities P_i, can be expressed in terms of Shannon's entropy (Shannon 1948; Shannon and Weaver 1949):

$$H(P) = - \sum_{i=1}^{n} P_i \log_b P_i \qquad (3.29)$$

The units in which the uncertainty is expressed are determined by the base b of the logarithm in this equation. For the *binary logarithm*, **1b**, $b = 2$, whose use is generally recommended by IUPAP for information, the unit is a **bit** (for the natural logarithm, ln, $b = e$, the unit would be a nit). The use of binary logarithms has been introduced by Hartley (1928) so that for an alternative (two-value) phenomenon, where $P_1 + P_2 = 1$, $H(P) = 1$ bit at $P_1 = P_2 = \frac{1}{2}$. If $n > 2$, we have $0 \le H(P) \le \text{lb } n$.

Therefore in this case **relative entropy**

$$h(P) = \frac{H(P)}{\text{lb } n} = -\frac{1}{\text{lb } n} \sum_{i=1}^{n} P_j \text{ lb } P_j \qquad (3.30)$$

is occasionally used. The relation $0 \le h(P) \le 1$ is also satisfied.

We do not consider it appropriate, however, to distinguish between the terms "thermodynamic entropy" and "information entropy." Indeed "entropy" is a universally valid measure of uncertainty, disorder, irrespective of the system whose disorder is characterized by it.

The entropy characterizing the uncertainty of a continuous random quantity with a probability density $p(x) > 0$ for $x \in \langle x_1, x_2 \rangle$ is

$$H(p) = -\int_{x_1}^{x_2} p(x) \text{ lb } p(x) \, dx \qquad (3.31)$$

for $\int_{x_1}^{x_2} p(x) \, dx = 1$. This entropy acts as a measure of uncertainty of accurate (unbiased) direct measurements. Entropies for some continuous distributions are given in Table 3.1; for some other symmetrical distributions (Simpson's, Laplace's, and Student's distribution for various degrees of freedom), see Tarabcik (1992).

Table 3.1 Entropies of Some Continuous Functions (in Bits)

Distribution	Probability Density Function	Entropy
Uniform $U(x_1, x_2)$	Eq. (3.18)	$\text{lb}(x_2 - x_1) = \text{lb}(\sigma\sqrt{12})$
Normal $N(\mu, \sigma^2)$	Eq. (3.21)	$\text{lb}(\sigma\sqrt{2\pi e})$
Truncated normal $TN(\mu, \sigma^2, x_0)$	Eq. (3.28)	$\text{lb}\{[1 - \phi(z_0)]\sigma\sqrt{2\pi e}\}$ $+ \frac{1}{2}z_0\varphi(z_0)/[1 - \phi(z_0)]$
Lognormal $LN(\mu, \sigma^2; x_0)$	Eq. (3.27)	$\text{lb}(\mu\sigma\sqrt{2\pi e})$
$\sqrt{12} = 3.464$	$\sqrt{2\pi e} = 4.133.$	

Table 3.2 Various Applications of Uncertainty and Measures of Information Gain

Quantity	Uncertainty	Equation	Information Content (Information Gain)	Equation
Nominal Ordinal — Discrete	$H(P_0) = -\sum P_{0i}\,\text{lb}\,P_{0i}$ $H(P) = -\sum P_i\,\text{lb}\,P_i$	(3.29)	$I = H(P_0) - H(P)$	(3.34)
Cardinal — Continuous	$H(p, p_0) = -\int p(x)\text{lb}\,p_0(x)\,dx$ $H(p, p) = -\int p(x)\text{lb}\,p(x)\,dx$	(3.31)	$I(p, p_0) = H(p, p_0) - H(p, p)$ Unbiased results	(3.41)
	$H(r, p_0) = -\int r(x)\text{lb}\,p_0(x)\,dx$ $H(r, p) = -\int r(x)\text{lb}\,p(x)\,dx$	(3.32)	$I(r, p, p_0) = H(r, p_0) - H(r, p)$ Biased Results	(3.42)

Since quantitative results can carry a mean error $\delta_i = x_i - \mu_i \neq 0$, and a priori uncertainty cannot be assumed to be based on an absolutely correct assumption or on absolutely correct preliminary information, a more adequate way to express uncertainty is in terms of Kerridge's (1961) and Bongard's (1966) *inaccuracy measure*:

$$H(r;p) = H(r) - D(r;p) = -\int_{x_1}^{x_2} r(x) \, \text{lb} \, p(x) \, dx \quad (3.32)$$

where the "error term"

$$D(r;p) = \int_{x_1}^{x_2} r(x) \, \text{lb} \, \frac{r(x)}{p(x)} \, dx \quad (3.33)$$

expresses the dissimilarity, divergence of the accurate distribution $r(x)$, and the experimental distribution $p(x)$ (see the monographs by Eckschlager and Stepánek 1979, 1985, and Table 3.2). Note that $r(x) = p(x)$ turns $H(r, p)$ into $H(p, p)$, which equals to Shannon's entropy $H(p, p) = H(p)$.

3.7 INFORMATION CONTENT

The information content of a signal, of a result of observation and of an accurate, unbiased result of simple direct measurement, can be expressed as the difference of Shannon's entropies,

$$I = H(P_0) - H(P) \quad (3.34)$$

for the discrete case of the a priori P_0 and a posteriori P probabilities or conditional probabilities, and as

$$I = H(p_0) - H(p) \quad (3.35)$$

for the case of the continuous a priori $p_0(x)$ and a posteriori $p(x)$ distributions.

For the same a priori and a posteriori distributions—for instance, for the a priori normal distribution $N(\mu_0, \sigma_0^2)$ and a posteriori normal distribution $N(\mu, \sigma^2)$, $\sigma^2 \leq \sigma_0^2$— we have

$$I = \text{lb}\frac{\sigma_0}{\sigma} \qquad (3.36)$$

For a priori uniform distribution $U(x_1, x_2)$ and a posteriori normal distribution $N(\mu, \sigma^2)$, we have

$$I = \text{lb}\frac{x_2 - x_1}{\sigma\sqrt{(2\pi e)}} \qquad (3.37)$$

The term in the numerator expresses the uncertainty about the result before the measurement. It represents the **expectation range** of the result. In the general case the expectation range is given by the upper x_2 and lower x_1 limits of the measuring range. The term in the denominator expresses the remained uncertainty about the result after the measurement; it is represented by the **confidence interval** $\Delta x = \sigma \sqrt{(2\pi e)}$. The facts are illustrated in Fig. 3.5.

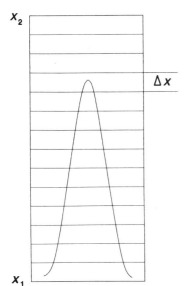

Figure 3.5. Expectation range and confidence interval of an analytical signal

Information content can be expressed by means of variance reduction (see Kateman 1986):

$$R = \left(\frac{\sigma}{\sigma_0}\right)^2 \qquad R \in \langle 0, 1 \rangle \qquad (3.38)$$

as

$$I = -\tfrac{1}{2} \ln R \qquad (3.39)$$

From this standpoint, using the terms given in the first and second rows of Table 3.2, Eq. (3.37) can be expressed as

$$I = \text{lb} \frac{\sigma_0 \sqrt{12}}{\sigma \sqrt{(2\pi e)}} \qquad (3.40)$$

Variance reduction is more important theoretically than in practice. In actual analytical practice a priori variances σ_0^2 are seldom known. So, in place of them, an expectation range is specified. More on the information content concept can be found in the monograph by Eckschlager and Stepánek (1985).

3.8 INFORMATION GAIN AND TRANSINFORMATION

In determining the information gain of analytical results when we have a priori uncertainty by means of the Kerridge-Bongard measure, we use the **divergence measure** (Eckschlager 1975; Eckschlager and Stepánek 1979):

$$I(p, p_0) = H(p; p_0) - H(p, p) = \int_{x_1}^{x_2} p(x) \, \text{lb} \frac{p(x)}{p_0(x)} \, dx \quad (3.41)$$

The divergence measure assumes that the results are accurate and unbiased. When the results are accurate, the a posteriori uncertainty is expressed by means of Shannon's entropy, and $r(x) = p(x)$ is inserted in the Kerridge-Bongard measure, which expresses the a priori uncertainty. When the quantitative result is not accurate, both the a priori and a posteriori uncertainties are expressed by means of the Kerridge-Bongard measure of inaccuracy, and the information gain is expressed as

$$I(r; p, p_0) = H(r; p_0) - H(r; p) = \int_{x_1}^{x_2} r(x) \, \text{lb} \frac{p(x)}{p_0(x)} \, dx \quad (3.42)$$

This measure in fact is the most general expression for the information gain of quantitative results; we refer to it as the **extended divergence measure** (Eckschlager and Stepánek 1985); see Table 3.3. Its usefulness lies in the fact that the actual distribution $r(x)$ does not need to be known; knowledge of its expected value and of its variance are sufficient.

The information content of analytical signals in Table 3.3 is used to assess simple measurements such as the intensity of the signal, but information gain is used to assess analytical results. It is clear that between the information content of the signal according to Eq. (3.35) and the information gain according to Eq. (3.41) or (3.42) there exists a dependence, which is given by the way the signal is decoded. In general, this relation can be regarded as a binary relation of sets of signals and results, as shown in Fig. 3.6. The perfectness of transformation of information about the signal into analytical information is characterized by means of **transinformation**.

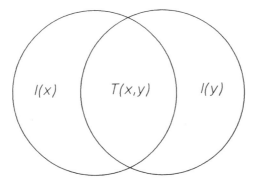

Figure 3.6. Venn diagram of the transinformation $T(x, y)$ between the sets of the information gain of results $I(x)$ and the information content of analytical signals $I(y)$

The transformation measure of mutual information is a particular case of the divergence measure, Eq. (3.41), and it characterizes information on ξ contained in η as

$$T(x, y) = \iint_{xy} p(x, y) \, \text{lb} \frac{p(x, y)}{p(x)p(y)} \, dx \, dy \qquad (3.43)$$

Transinformation is a symmetrical quantity, $T(x, y) = T(y, x)$, and for two equal random quantities it passes into entropy, $T(x, x) = H(p(x))$. When the two random quantities ξ and η do not mutually correlate (i.e., neither contains information on the other), the transinformation is zero. Transinformation was introduced to analytical chemistry by Frank et al. in 1982. But, to a first approximation, mutual information can be characterized by the correlation coefficient r_{xy} or by the determination coefficient $D = 100r_{xy}^2$: The closer that r_{xy} approaches 1 (and D approaches 100), the better is the mutual information between the two random variables:

$$T(x, y) = \text{lb} \frac{1}{\sqrt{1 - r_{xy}^2}} = -\text{lb} \sqrt{1 - \frac{D}{100}} \qquad (3.44)$$

Details on the correlation coefficient can be found in any mathematical statistics textbook; see also the monograph by Eckschlager (1969, ch. 5.4).

REFERENCES

Bongard, M. M. 1966. *Problemy Kibernetiki* **6**, 91 (in Russian).

Eckschlager, K. 1969. *Errors, Measurement and Results in Chemical Analysis*. Van Nostrand, London.

Eckschlager, K. 1975. *Fresenius Z. Anal. Chem.* **277**, 1.

Eckschlager, K., and Stepánek, V. 1979. *Information Theory as Applied to Chemical Analysis*. Wiley, New York.

Eckschlager, K., and Stepánek, V. 1985. *Analytical Measurements and Information. Advances in the Information Theoretic Approach to Chemical Analyses*. Research Studies Press, Letchworth.

Frank, I., Veress, G., and Pungor, E. 1982. *Hung. Sci. Instr.* **54**, 1.

Hartley, R. V. L. 1928. *Bell. Syst. Techn. J.* **7**, 535.

Kateman, G. 1986. *Anal. Chim. Acta* **191**, 215.

Kerridge, D. F. 1961. *J. Roy. Statist. Soc.* Ser. B, **23**, 184.

Malissa, H., Rendl, J., and Marr, I. L. 1975. *Talanta* **22**, 597.

Shannon, C. E. 1948. *Bell Syst. Techn. J.* **27**, 379, 623.

Shannon, C. E., and Weaver, W. 1949. *The Mathematical Theory of Communication*. University of Illinois Press, Urbana.

Tarabcik, P. 1992. *Chem. Listy* **86**, 648.

Vajda, I., and Eckschlager, K. 1980. *Kybernetika* **16**, 120.

CHAPTER

4

IDENTIFICATION OF COMPONENTS

Both identification and qualitative analysis deal with determining what type of analytes are contained in a sample. While **qualitative analysis** seeks to answer the question of whether certain components are present in the sample, **identification analysis** aims to answer the question of what unknown substance is the signal. Stated in other words, in qualitative analysis the signal sought is in a position z_j which is known beforehand, whereas in identification analysis one seeks to find to which substance the observed signals belongs. Qualitative and identification chemical analysis requires information-theoretic evaluation. Unlike the situation in quantitative analysis, statistical methods are less useful.

As a rule, identification analysis of elementary components (elements, ions) uses the same methods as qualitative analysis but proceeds by other strategies. Frequently identifications are carried out by a sequence of qualitative-analytical decisions. Identification of compounds mostly combines several methods and procedures, some of which—instrumental methods in particular—are closest to the methods of structural analysis and separation.

4.1 METHODS OF IDENTIFICATION ANALYSIS

Chemical reactions as well as physical interactions are suitable to identify chemical components—namely elements, ions, compounds, or parts of them (functional groups, fragments, and so on) which can be reconstituted afterward like a puzzle. Figure 4.1 gives an overview of identification strategies in analytical chemistry. These strategies are independent of the kind of chemical sample and analyte (organic or inorganic).

The measurement of a **characteristic physical property** is only suitable for the identification of pure substances. Physical proper-

45

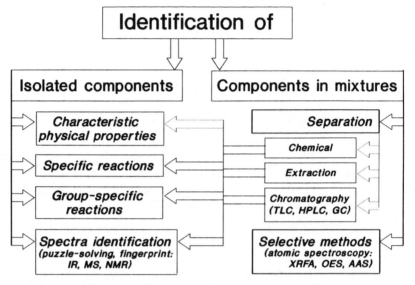

Figure 4.1. Strategies and methods of separation

ties like density, refraction index, melting point, and other nonspecific quantities can be measured with high precision and therefore
are indicative of a pure substance. **Specific chemical reactions** are
typical for only one component under certain conditions. **Specifity**
is a characteristic that was defined for the first time by Kaiser
(1972) on the basis of the **matrix of partial sensitivities.**

$$
\Gamma = \begin{pmatrix}
\gamma_{11} & \gamma_{12} & \cdots & \gamma_{1n} \\
\gamma_{21} & \gamma_{22} & \cdots & \gamma_{2n} \\
\vdots & \vdots & & \vdots \\
\gamma_{n1} & \gamma_{n2} & \cdots & \gamma_{nn}
\end{pmatrix}
\tag{4.1}
$$

where $\gamma_{ik} = \partial y_i / \partial x_k$ is the partial sensitivity, namely the change of
the measuring quantity y (e.g., the absorptivity at a fixed wave-
length—in identification analysis possibly detected only by a given
color) at its characteristic value of the analyte i by a certain amount
x of the analyte k. It is evident that the partial sensitivity for $i = k$
passes to the general sensitivity $S_i = \gamma_{ii} = \partial y_i / \partial x_i$.

The definition of specifity Ψ_a (Kaiser 1972) is expressed as

$$\Psi_a = \frac{|\gamma_{aa}|}{\sum_{k=1}^{n}|\gamma_{kk}| - |\gamma_{aa}|} - 1 \qquad (4.2)$$

Complete specifity ($\Psi_a \to \infty$) related to a certain component a can be stated if only the one element γ_{aa} of the diagonal of the matrix is different from zero.

In analytical work specific reactions are carried out on the basis of typical precipitates and their behavior (e.g., AgCl, Ni-diacetyl-dioxim), color reactions (e.g., $Fe(SCN)_3$), crystal shapes (e.g., the characteristic snow crystals of $NH_4MgPO_4 \cdot 6H_2O$), or flame coloring. As a rule, specific reactions are performed by micro techniques, particularly as spot tests. **Group-specific chemical reactions** identify components whatever the number present. By combining several of group-specific reactions and coupling them in a fixed logical sequence, it is possible to determine the components step by step, for example, **systematic separation processes** (cation- and anion-separation schemes). Figure 4.2 shows the standard scheme for the identification of cations. Another possibility of group-specific reactions is represented by derivative reactions obtained before characteristic physical properties such as the melting point are measured.

Quantitative analysis may also be used in identifying components, especially of organic compounds, **elemental analysis** (CHN analyzers). However, the results are frequently insufficient and require additional information, such as the shape of the isotope pattern in normal-resolution mass spectrometry or precise mass values in high-resolution mass spectrometry.

Molecular spectroscopy is the most efficient tool for identifying compounds. As a rule, its application is limited to the identification of isolated components. In some cases multiple component identification may be carried out, or contaminations and by-products recognized. The identification process differs according to method. Among the most important are the following:

1. Vibration spectroscopy (infrared and Raman spectroscopy).
2. Nuclear magnetic resonance (NMR) spectroscopy.
3. Mass spectrometry.

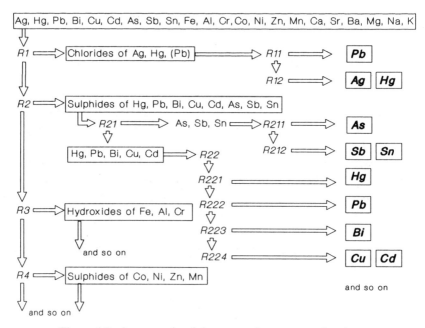

Figure 4.2. An example of the separation process of cations

By comparison, UV/Vis spectroscopy can be singled out as exceptionally able to identify components without additional information from other methods. But, as a rule, information of more than one method is needed to recognize any compound. Without recourse to a direct comparison with spectra as fingerprints, the identification analysis proceeds as a puzzle (Danzer, Than, Molch, and Küchler 1987). Vibration spectroscopy gives information about the functional groups of a molecule. The rules on how to connect these "building blocks" are disclosed by neighboring relations which are derived mainly from signal shift and signal multiplicity obtained by NMR spectroscopy. Additional conclusions about the composition of a molecule can be drawn from fragmentation analyses in mass spectrometry.

Components in mixtures are mainly identified by separation procedures. The most important are chromatographic methods. Chromatographic methods are usually used to identify organic compounds, but by ion chromatography inorganic systems can be identified as well. Depending on the separation system, which has to be optimized for the stationary and mobile phases as well as for operation conditions, a complete separation of the components of a chemical composition may be obtained. The most important chromatographic methods in qualitative analysis are three:

1. Gas chromatography (GC).
2. High-performance liquid chromatography (HPLC).
3. Thin-layer chromatography (TLC).

Identification of components is carried out either by direct comparison with reference standards for substances—by parallel runs or additional runs—or by isolation, and then analysis by the procedures mentioned above.

Separation methods teamed up with spectroscopic identification methods have increasingly enabled successful environmental, biochemical, and pharmaceutical investigations. Called **hyphenated methods**, they are used to identify all types of compounds, organic and inorganic, elemental as well as complex species (e.g., GC–MS, GC–FT–IR, HPLC–MS, MS–MS, MS^n, GC–AAS, HPLC–AAS, HPLC–ICP).

The most important feature that spectroscopic methods offer for the identification of chemical components is selectivity; that is, through spectroscopy components of chemical mixtures can be detected independent of each other. Our definition of selectivity Ξ comes from Kaiser (1972) who derived it from the matrix of partial sensitivities (Eq. 4.1) as

$$\Xi = \min_{i=1\ldots n} \left(\frac{\gamma_{ii}}{\sum_{i=1}^{n}|\gamma_{ik}| - |\gamma_{ii}|} - 1 \right) \qquad (4.3)$$

Ideally an analytical method is selective if all nondiagonal elements γ_{ik} of the matrix (4.1) are zero (excluded are diagonal elements γ_{ii}). Each component can then be analyzed independently of the others. It is evident that the better this condition is fulfilled, the higher is the resolution power of the analytical method. By Eq. (4.3) a given degree of selectivity can be expressed: larger values of Ξ mean that there is a high selectivity with the analytical method

($\Xi \to \infty$ for complete selectivity), whereas small values indicate low selectivity.

Selectivity can be characterized by a quantity $H(\gamma_{ik})$ which is analogous to the entropy computed from the matrix of partial sensitivities (4.1): $-\Sigma_i \Sigma_k \gamma_{ik}$ lb γ_{ik}. In the ideal case where selective detection occurs for all analytes, $H(\gamma_{ik}) = 0$ (for a discussion see Eckschlager 1991).

Atomic-spectroscopic methods are mainly used to identify elements. Mass spectroscopy may be considered where there is low resolution because of coinciding signals which may be influenced by the isotope pattern of the element. In most cases, however, such coincidences can be easily accounted for by computation.

To point out the distinction between the identification analysis of unknown elements and the qualitative analysis of selected elements, we turn to the example of atomic emission spectrography. Although, in principle, the analytical procedure and the recorded spectra are the same, some differences exist in the evaluation strategies. In an *identification* analysis one searches the spectrum for characteristic lines—mainly those strong intensity. Then, by consulting a wavelength-ordered table or an atlas, one can identify the elements present in the sample. In *qualitative analysis* one checks the spectrum to see if the line of a certain element is present or absent. If necessary, the examination is extended to tables of spectra wavelengths ordered according to the elements.

The most important methods for determining the elements—and in some cases their speciation as indicated by bonding and oxidation states, valency, and coordination—are the following:

1. Mass spectroscopy (excitation by spark, ICP, and glow discharge, GD-MS).
2. Atomic emission spectroscopy (excitation by flame, ICP, arc, spark; GD-OES).
3. X-ray fluorescence and X-ray emission spectroscopy.
4. Auger- and photoelectron spectroscopy (AES, XPS, UPS, ESCA).
5. Mößbauer spectroscopy.

The information gain of identification analyses is dependent on the analytical problem and the sample (the matrix and the kind and number of components to be identified) as well as on the method used and the analytical strategy.

4.2   INFORMATION UNCERTAINTY AND INFORMATION CONTENT OF IDENTIFICATION RESULTS

In identification analysis the decisive factor is the **position** of a signal's peak correlated with its nonzero intensity; in other words, the identity is assigned according to the position of the peak. The relationship between the input and output consists in assigning a position to a certain substance. As identification results are being analyzed, one must consider the way the information is received by the signal decoding feature. In general, there are two ways in which the connection between analytes and signals can be described. Figure 4.3 shows both empirical and systematical assignment effects.

Empirical connections are represented by tables or an atlas (e.g., in IR, NMR, optical atomic spectroscopy); systematical connections are described by natural laws (e.g., the Moseley law) of the dependency of wavelengths (e.g., of the K_α lines) from the atomic number. Information gain expressed in terms of the divergence measure is different from the information content of analytical results.

Let us first deal with the information content of identification results. In the simplest case, where one component is to be identified out of a number of n_0, its presence can be presumed with the probability. $P(A_i) = 1/n_0$ $(i = 1, 2, \ldots, n_0)$ and Shannon's entropy $H(P) = \text{lb } n_0$. The information content of identification is then

$$I(1, n_0) = \text{lb } n_0 - \text{lb } 1 = \text{lb } n_0 \qquad (4.4)$$

where n_0 is the number of possible components, presumed prior to performing the experiment, and one component is identified with safety. For the ambiguous case where after the experiment no component can be distinguished between n components, the information content is

$$I(n, n_0) = \text{lb } n_0 - \text{lb } n = \text{lb } \frac{n_0}{n} \qquad (4.5)$$

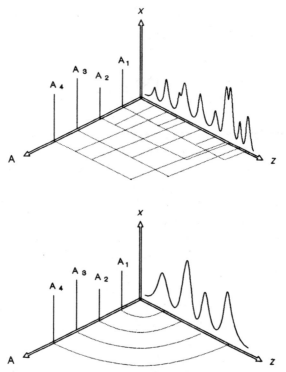

Figure 4.3. Different ways in which analytes can be assigned to signals: empirical (top) and theoretical (bottom)

The dependence of the information content I on lb n_0 according to Eq. (4.5) for several values of n is shown in Fig. 4.4. Equation (4.5) is an expression for the information content of an experiment, so it is not confined just to identification analysis; it has much broader application.

The ratio $n/n_0 \in \langle 0, 1 \rangle$ can be used to reduce the number of possible variants from n_0 to $n < n_0$ following the experiment or evaluation. The results of **chemometrics** (e.g., pattern recognition by *classification* or *cluster analysis*) can be evaluated in the same way. Often the only way to express the information content of an experiment is by the reduction of the number of variants. The result provides a *nominal* or *ordinal* quantity.

In the case where the components cannot all be expected to have the same a priori probability, the information content in the identi-

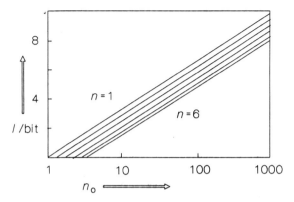

Figure 4.4. Information content of identification analysis

fication analysis is expressed in terms of the difference between Shannon's entropy:

$$I = H_0(P) - H(P) = -\sum_{i=1}^{n_0} P_0(A_i)\, \text{lb}\, P_0(A_i)$$

$$+ \sum_{i=1}^{n} P(A_i)\, \text{lb}\, P(A_i) \qquad (4.6)$$

When we calculate the a posteriori entropy for $P(A_i) = 0$ (i.e., to rule out the presence of the ith analyte), then we insert $0\, \text{lb}\, 0 = 0$. For the unambiguous case where the identification of one component is involved, we have $I = H_0(P)$.

When there are q components out of a number n_0 that are to be identified simultaneously or sequentially, we proceed in a different way from the case of one component of n_0 limited between that and n (Eq. 4.5). The **information amount** $M(q; n_0)$ of identification analysis of q components is given by the sum of the information contents of each of the q_j components:

$$M(q; n_0) = \sum_{j=1}^{q} I(1, n_{0,j}) \qquad (4.7)$$

The characteristic of such identifications is that the number of $n_{0,j}$

in each step reduces by one, according to $n_{0,j} = n_0, (n_0 - 1), \ldots,$ $(n_0 - q_{j-1})$ for $j = 1 \ldots q$. If, for example, three components out of a possible number of 100 are to be identified, $M(3; 100) = I(1; 100) + I(1; 99) + I(1; 98)$. By Eq. (4.5) this results in $M(3; 100) = (6.644 + 6.629 + 6.615)\,\text{bit} = 19.89\,\text{bit}$.

When $n_0 \gg q$, and the information content of each of the q components is equal, Eq. (4.7) can be simplified to

$$M(q; n_0) = q \cdot I(1; n_0) \tag{4.8}$$

Using Eq. (4.8), we obtain $M(3; 100) = 19.932\,\text{bit}$. Such multiple-component analyses would take place in studies of environmental organic and toxicological pollutants where the number of compounds in question can run into thousands or hundreds of thousands.

When the number q of actual components is unknown, the identification process consists of two steps:

1. Find the number q_a of actual components out of a number n_0 contained in the substance under investigation (e.g., by using chromatographic procedures). This case is described by Eqs. (4.5) and (4.6), but here the information content is expressed as $I(q_a, n_0) = \text{lb}(n_0/q)$.
2. Identify the q components by single or by combined methods (e.g., GC–MS). The relevant expression for the information amount is Eq. (4.5).

Then the total information amount of the complete identification process is expressed as

$$M(q_a; n_0) = I(q_a, n_0) + M(q; n_0)$$
$$= \text{lb}\frac{n_0}{q} + \sum_{j=1}^{q} I(1, n_{0,j}) \tag{4.9}$$

Suppose that we are faced with the problem of identifying the unknown toxic components in a given sample known to have caused poisoning. This is typical of the "general unknown" cases in toxicology. The number of components n_0 may be as large as 10^5 in this case. Say, we recognize by means of GC that $q = 7$ components, which can then be determined by identification analysis (or by parallel analysis GC–MS). The information amount would be $M(q_a, 10^5) = \mathrm{lb}(10^5/7) + 7 \cdot \mathrm{lb}\, 10^5 = (13.80 + 116.27)\,\mathrm{bit} = 130.07\,\mathrm{bit}$ provided that the information contents of all the compounds are maximum and equal.

Since Eqs. (4.4) through (4.9) express the information content of only the experiment, we must use a divergence measure for the actual information gain of the identification result:

$$I(P, P_0) = \sum_{i=1}^{n} P(A_i)\, \mathrm{lb}\, \frac{P(A_i)}{P_0(A_i)} \qquad (4.10)$$

In an instrumental identification analysis, where signals can overlap and/or their positions can be determined with only a limited precision or are uncertain, the information content of the result can be obtained by using entropy for conditional probability. Denote $P(z_j|A_i)$ the conditional probability that the signal appears in position z_j, $j = 1,\ldots,m$, if analyte A_i, $i = 1,\ldots,n$, is present. This probability can be easily determined experimentally; in fact $P(z_j|A_i)$ is usually either unity (signal occurs) or zero (signal does not occur).

The probabilities $P(A_i|z_{ji})$ necessary to decode the signal can be found by using Bayes's relation, Eq. (3.6). The uncertainty of the identification can be expressed through **compound entropy**, which in instrumental identification analysis is written

$$H\big(P(A_i|z_{ji})\big) = - \sum_{i=1}^{n} P(A_i|z_{ji})\, \mathrm{lb}\, P(A_i|z_{ji}) \qquad (4.11)$$

The compound entropy function is similar to that for the selectivity of detection. The conditional probability $P(A_i|z_j)$ can be inserted for the a posteriori probability when solving for the divergence measure.

The difference between entropies I, Eq. (3.34), expresses only the information content of the qualitative detection. The actual information gain (i.e., the interpretation of experimental result) can be expressed using the divergence measure $I(P, P_0)$, Eq. (4.10). For unambiguous identification we have $H(P) = 0$, and thus $I = H_0 = \sum_i P_0(A_i) \mathrm{lb}\{1/[P_0(A_i)]\}$ and $I(P, P_0) = \mathrm{lb}\{1/[P_0(A_i)]\}$, since we have set $0 \, \mathrm{lb} \, 0 = 0$; see Eq. (4.6). This information gain, which is important for identification analysis, was designated "specific information content," $I_{sp}(A_i)$, by Meyer-Eppler (1969); see Eq. (4.12).

The meaning of the two quantities I and $I(P, P_0) = I_{sp}(A_i)$ will be illustrated by a simple example. Assume that two cations, lead and cadmium, may be present ($n_0 = 2$) and that their presence is equally probable. The information content of the H_2S assay then is $I = 1$ bit irrespective of its result. However, if cadmium is expected with a probability $P(Cd) = 0.2$ and lead is expected with a probability $P(Pb) = 0.8$, then the information content is invariably $I = H_0 = -(0.8 \, \mathrm{lb} \, 0.8 + 0.2 \, \mathrm{lb} \, 0.2) = 0.72 \, \mathrm{bit}$. On the other hand, the information gain for the presence of lead which we a priori expected is evidenced with a higher probability, $I_{Pb}(P, P_0) = 1 \, \mathrm{lb}(1/0.8) = 0.32 \, \mathrm{bit}$, and, on the other, to our surprise, cadmium is detected, with an information gain of $I_{Cd}(P, P_0) = 1 \, \mathrm{lb}(1/0.2) = 2.32 \, \mathrm{bit}$. Unlike the information content, the information gain, which obtains as a specific measure for every component, allows for a more or less expected result. The information content provides only a general uncertainty reduction.

4.3   PRACTICAL APPLICATIONS

The first example comes from the laboratory training of our advanced students. One of the tasks is to identify the components in a number of metals, among whose samples are alloys and modified metalloids and ores. The students are asked to select a method that will determine the components of the materials.

We are interested in two cases. In the first, the sample is from a bright, silvery metallic object; the students have no a priori information about its constituents. For the components nearly all the elements of the periodic table may be considered, with different probabilities for their presence. Let us suppose that $n_0 = 70$ is the number of possible elements. From the student's experience,

$n_{0,1} = 20$ (e.g., Mg, Al, Si, Ti, V, Cr, Mn, Fe, Co, Ni, Zn, Ge, Mo, Ag, Sn, Sb, W, Pt, Pb, Bi) can be expected with a higher probability, $n_{0,2} = 30$ with a normal, and $n_{0,3} = 20$ (e.g., the rare earth and the alkali elements) with a less probability. These facts will be considered by weighting factors: $w_1 = 1.5$, $w_2 = 1.0$, and $w_3 = 0.5$, with the application $w_1 n_{0,1} + w_2 n_{0,2} + w_3 n_{0,3} = n_0$. Therefore the probabilities for the three groups of elements are $P(A_{i,1}) = 1.5/70 = 0.0214$, $P(A_{i,2}) = 1/70 = 0.0143$, $P(A_{i,3}) = 0.5/70 = 0.00714$. The total a priori entropy, according to Eq. (4.6), is

$$H_0(P) = 20\left(0.0214\,\mathrm{lb}\,\frac{70}{1.5}\right) + 30[0.0143\,\mathrm{lb}(70)]$$

$$+ 20[0.00714\,\mathrm{lb}(70/0.5)]$$

$$= (2.37 + 2.63 + 1.02)\,\mathrm{bit} = 6.02\,\mathrm{bit}$$

Now we assume that by the identification analysis the students have found that Cr and Si are the major constituents and Fe the minor one (e.g., as in a resistor material). Because the identification analysis was carried out by OES using a large grating spectrograph (PGS-2, Carl Zeiss Jena, resolution: 45,600; dispersion: 0.71 nm/ mm) and the identification of lines and thus of elements could be made unambiguously, uncertainties of instrumental identification according to Eq. (4.10) can be neglected.

The information content of the identification in this case is therefore approximately, from Eq. (4.8), $M(3; 70) \approx 3\,H_0(P) = 18.06\,\mathrm{bit}$. An exact calculation can be carried out using Eq. (4.7): $M(3; 70) = I(1; 70) + I(1; 69) + I(1; 68)$. The actual probabilities are as follows:

$I(1; 70) = 6.02$ bit (see above).
$I(1; 69) = 19[0.0217\,\mathrm{lb}(69/1.5)] + 30[0.0145\,\mathrm{lb}\,69]$
$\qquad + 20[0.00725\,\mathrm{lb}(69/0.5)] = 5.96$ bit.
$I(1; 68) = 18[0.0221\,\mathrm{lb}(68/1.5)] + 30[0.0147\,\mathrm{lb}\,68]$
$\qquad + 20[0.00735\,\mathrm{lb}(68/0.5)] = 5.91$ bit.

In this way we obtain the precise information amount of $M(3; 70) =$ 17.89 bit, which is not very different from the approximate value given earlier.

In the second case, the sample is a small piece of a bright yellow material. The students might guess that Cu, Zn, Sn, Au, and Ag can be found with somewhat higher probability than the other elements, and the other assumptions might be similar to those of the first case. Proceeding from this, the following probability relations R_i may be expected for the elements:

Cu : Zn : Au : Ag : Sn : {elements of group 1}[1]
12 :12: 8: 8: 5:   1.5

: {elements of group 2}[2] : {elements of group 3}[2]
:   1.0   :   0.5

From these relationships the following probabilities can be obtained in terms of $P(A_i) = R_i/\Sigma R_i$:

$$P(\text{Cu}) = P(\text{Zn}) = \frac{12}{107.5} = 0.1116$$

$$P(\text{Au}) = P(\text{Ag}) = 0.0744$$

$$P(\text{Sn}) = 0.0465$$

$$P(A_{i,1}) = 0.01395$$

$$P(A_{i,2}) = 0.00930$$

$$P(A_{i,3}) = 0.00465$$

By Eq. (4.6) we have the total a priori entropy results:

$$H_0(P) = -2(0.1116 \text{ lb } 0.1116 + 0.0744 \text{ lb } 0.0744)$$
$$- 0.0465 \text{ lb } 0.0465 - 15(0.01395 \text{ lb } 0.01395)$$
$$- 30(0.0093 \text{ lb } 0.0093) - 20(0.00465 \text{ lb } 0.00465)$$
$$= 0.7061 + 0.5578 + 0.2058 + 1.2898 + 1.8829 + 0.7206$$
$$= 5.36 \text{ bit}$$

[1] As classified in the first example minus the five selected elements, $n_{0,1} = 15$.
[2] As classified in the first example, $n_{0,2} = 30$ and $n_{0,3} = 20$.

If the students find Cu and Zn to be the major constituents and Si the minor constituent, the information amount of the identification is approximately $M(3; 70) \approx 3H_0(P) = 16.08$ bit, by Eq. (4.8); the exact value according to Eq. (4.7) would be somewhat less). Compared with that of the first example where $M(3; 70) \approx 18.06$ bit was obtained, this is less information. Clearly a priori the color gives an indication of what is present in the material. Yet this example suggests various other possibilities since, besides brass, bronze, and alloyed gold, other yellow-colored substances that might be considered are titanium nitride, pyrite, and some ores.

Quite another problem may be considered if a similar sample—a piece of a yellow sheet metal—were to be identified by a laboratory that routinely deals with the assay of precious metals. The expectations, expressed by a priori probabilities, may be $P(Cu + Zn) = 0.75$, $P(Au) = 0.20$, and $P(\text{other}) = 0.05$.

From this, the following a priori entropy results are obtained:

$$H_0(P) = -2(0.75)\,\text{lb}(0.75) - 0.2\,\text{lb}\,0.2 - 0.05\,\text{lb}\,0.05 = 1.30\,\text{bit}$$

We now have less information entropy than in the last example, but the result reflects the work of a more experienced analyst in the field. The average information content for each element is about 1.30 bit, which is independent of other results. A more precise identification analysis can be made by using the specific information gain function $I_{sp}(A_i)$ of a given result A_i (Meyer-Eppler 1969):

$$I_{sp}(A_i) = -\text{lb}\,P_{Ai} \qquad (4.12)$$

This function yields the following specific information gains:

$I(Cu) = I(Zn) = 0.415$ bit,
$I(Au) = 2.322$ bit,
$I(\text{other}) = 4.322$ bit.

Clearly there is an element of surprise in obtaining specific information gain for certain elements.

Another example, which we will only sketch out roughly, involves complex environmental analysis, which is an important area of analytical practice. Impairment of water quality may have any number of causes and may concern organic or inorganic pollutants. So the identification strategy would need to proceed stepwise:

1. Collect preliminary information on the system and its contaminants.
2. Determine summary and group-specific parameters, for example, TOC (total organic carbon), DOC (dissolved organic carbon), COD (chemical oxygen demand), BOD (biological oxygen demand), AOX (adsorbable organic halogen), AOS (adsorbable organic sulphur), SAC_{254} (spectral absorption coefficient at 254 nm).
3. Enrich analytes by extraction (liquid or solid phase) and/or adsorption.
4. Screen related classes of substances (PCAs, LHHCs, BTXs, PCBs, dioxines, pesticides, etc.).
5. Make single component identifications.

In the most cases of analytical environmental practice, the last step consists in a quantitative analysis of the components found.

In a similar detailed way the information content has to be calculated. All the steps of limiting of the substance groups and compounds classes are to be considered to the single components. Because all group-specific and single-component identifications need instrumental methods such as TLC, HPLC, and GC (mostly in their high resolution applications, capillary GC, and coupled with MS or selective ion monitoring) the divergence measure, Eq. (4.10), is important here. Both the chromatographic separations and the instruments will show some uncertainties and overlapping peaks.

For example, in a general unknown environmental identification problem, a rough estimate of the information amount might give the following: Let the number of components be $n_0 = 5000$. We want to reduce this number stepwise to 1000 (e.g., to halogen-containing compounds) and afterward to about 200 (e.g., PCBs). After subjecting a sample material GC/MSD (capillary gas chromatography coupled with a mass selective detector), we find that 60 are present and can be identified.

Using Eq. (4.8), we obtain an approximate information amount $M(60; 5000) \approx \text{lb}(5000/60) + 60\,\text{lb}\,5000 = 743.66\,\text{bit}$, which is independent of the number of intermediate steps. For an exact calculation we would have to consider all preliminary information such as causes of pollution (flue-ashes, incineration, pesticides), characteristic properties of the sample (smell, taste), and effects that limit the number of compounds. Further we would consider typical pattern of the single components within each substance class (see Hutzinger et al. 1974; Ballschmiter et al. 1987) and stipulate different a priori probabilities for each of the compounds. Last, we would consider any uncertainties of signal positions and signal interpretation. It should be clear that in establishing these parameters of multicomponent analysis, the boundaries between identification analysis and qualitative multicomponent analysis are not always sharp; they can overlap.

REFERENCES

Ballschmiter, K., Schafer, W., and Buchert, H. 1987. *Fresenius Z. Anal. Chem.* **326**, 253.

Clej, P., and Dijkstra, A. 1979. *Fresenius Z. Anal. Chem.* **298**, 97.

Danzer, K. 1973. *Z. Chem.* **13**, 229.

Danzer, K., Than, E., Molch, D., and Küchler, L. 1987. Analytik— Systematischer Überblick. Akademische Verlagsgesellschaft Geest & Portig, Leipzig.

Eckschlager, K. 1975. *Fresenius Z. Anal. Chem.* **277**, 1.

Eckschlager, K. 1991. *Collect. Czech. Chem. Commun.* **56**, 505.

Eckschlager, K., and Král, M. 1984. *Collect. Czech. Chem. Commun.* **49**, 2342.

Eckschlager, K., and Stepánek, V. 1978. *Mikrochim. Acta* **I**, 107.

Eckschlager, K., and Stepánek, V. 1979. *Information Theory as Applied to Chemical Analysis*. Wiley, New York.

Eckschlager, K., and Stepánek, V. 1981. *Mikrochim. Acta* **II**, 143.

Eckschlager, K., and Stepánek, V. 1985. *Analytical Measurement and Information. Advances in the Information Theoretic Approach to Chemical Analysis*. Research Studies Press, Letchworth.

Eckschlager, K., Stepánek, V., and Danzer, K. 1990. *J. Chemometrics* **4**, 195.

Hutzinger, O., Safe, S., and Zitko, V. 1974. *The Chemistry of BCB's*. CRC Press, Cleveland, OH.

Kaiser, H. 1972. *Fresenius Z. Anal. Chem.* **260**, 252.

Meyer-Eppler, W. 1969. *Grundlagen und Anwendungen der Informationstheorie*. Springer-Verlag, Berlin.

Liteanu, C., and Rica, I. 1979a. *Statistical Theory and Methodology of Trace Analysis*. Ellis Horwood, Chichester.

Liteanu, C., and Rica, I. 1979b. *Anal. Chem.* **51**, 1986.

CHAPTER

5

QUALITATIVE ANALYSIS

In performing qualitative analysis, we seek to answer the question of whether or not a certain component is present in the sample. For this, we observe if there is a corresponding reaction that manifests itself in an expected change in a solution's color, in the formation of a heterogeneous phase (precipitate, gas evolving, etc.) or, if in an instrumental analysis, a signal occurs in a position z_j that is known beforehand.

In general, the first step of a chemical or an instrumental analysis is rather simple. What is involved is transformation of experimental results into information. However, in toxicological or environmental analysis cases, this transformation may require the use of a computer with specialized software.

In practice, there is no real distinction made between qualitative and quantitative analyses. Sometimes qualitative analysis is viewed as quantitative analysis with rough readout, see Fig. 5.1. In Section 6.5 on trace analysis we will demonstrate that there is a continual transition between qualitative, semi-quantitative, and quantitative analysis of low contents.

5.1 METHODS OF QUALITATIVE ANALYSIS

The problem of qualitative analysis is often less extensive than that of identification. As a rule, the objective is not to find any unknown components but to examine if certain well-known components are present or absent. There are fewer methods for qualitative analysis than for identification purposes. The molecular spectroscopic methods used in the identification of compounds are usually not applied in qualitative analysis. In cases where one has to check if a certain compound appears as a by-product of a given reaction, the com-

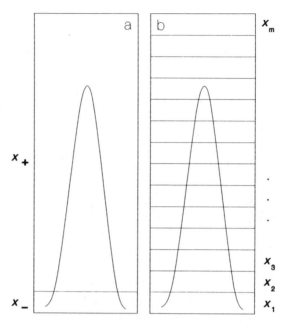

Figure 5.1. Signal detection in (*a*) qualitative and (*b*) quantitative analyses

pound may be detected as an undisturbed signal through a typical IR-, NMR- or mass spectrum of the major component.

The methods used in qualitative analysis must meet high standards in selectivity or specificity capabilities; see Eqs. (4.2) and (4.3). Therefore any method based on the measurement of a nonspecific property (density, melting point, and so on) such as is used in some cases for the identification of a component is not suitable for qualitative analysis. Rather, qualitative analysis is carried out mainly by methods of identification analysis, in particular by selective two-dimensional instrumental methods. Among the preferred methods of qualitative analysis are the following:

1. Chromatographic techniques for recognition of organic compounds (TLC, HPLC, GC) and of inorganic constituents (ion chromatography); the assay may be supported by comparison runs (parallel runs of reference substances or addition of them to the sample).

2. Methods of atomic emission spectroscopy and X-ray fluorescence for testing on chemical elements, if necessary also by means of comparison spectra.

3. Mass spectrometry of elements, isotopes, and species [vaporization and excitation by spark, arc, and plasma (ICP-MS)].

4. Photoelectron and AUGER electron spectroscopy for the analysis of elements where from typical chemical shifts valency and oxidation states may be recognized, and therefore qualitative analysis of species is possible.

5. Specific chemical reactions or reaction sequences of single or few components; chemical methods are simple as well as cheap and—in addition to some instrumental methods like RFA—well suited for analysis in field (e.g., in geological exploration or environmental screening).

Because the boundary between qualitative and quantitative analysis is indistinct, qualitative results are often expressed in quasi-quantitative terms without the proper precision. From practical experience any analyst can state whether a certain component is a major or minor constituent of the sample or only a trace component. This kind of quantification will be considered later in connection with quantitative analysis.

5.2 INFORMATION UNCERTAINTY AND INFORMATION CONTRIBUTION OF QUALITATIVE RESULTS

In qualitative analysis one still has to explain how input relates to output, as in identification analysis, as well as assign a visual chemical reaction or of a signal position to an analyte. But, as mentioned above, the content of the information differs from that in identification analysis. The question in *identification analysis* is, "Which constituent(s) is present in the sample," whereas in *qualitative analysis* we ask, "Is a given constituent(s) present in the sample: yes or no?"

Clearly in the simplest qualitative analysis where the object is to ascertain the presence of one component, the probability of the two possible results *yes* and *no* is equal. Therefore the information content of the detection is given by the simplest information-theoretic relation

$$I = \text{lb } 2 = 1 \text{ bit} \tag{5.1}$$

Where the component cannot be expected with the same probability, the information content of qualitative detection is expressed in terms of the difference between the SHANNON's entropies, namely $I = H_o(P) - H(P)$; see Eqs. (3.1), (3.29), and (4.6). For unambiguous detection we have $H(P) = 0$, and thus $I = H_o(P)$. If the a priori probabilities $P_+(A_i)$ and $P_-(A_i)$ are different, the average information content of the qualitative detection is

$$I = -P_+(A_i) \text{ lb } P_+(A_i) - P_-(A_i) \text{ lb } P_-(A_i) \tag{5.2}$$

Rewritten in terms of some specific information content [see Eq. (4.12); MEYER-EPPLER 1969], the expression becomes

$$I = P_+(A_i)I_+(A_i) + P_-(A_i)I_-(A_i) \tag{5.3}$$

In this way the normal (average) information content is the sum of the specific information contents each weighted with its a priori probability. Table 5.1 shows the specific and average information contents for qualitative analyses under different conditions.

As Table 5.1 shows, different a priori probabilities may correspond to the following practical situations: $P_+(A_i) = 0.5$: no a priori information on the analyte (e.g, in an unknown sample); $P_+(A_i) = 0.9$: the presence of the analyte is probable to some

Table 5.1. Specific and Average Information Contents of Qualitative Analyses with Different a priori Probabilities of an Analyte A_i Present (in Bits)

A priori Probabilities		Specific Information Content		Average Information Content
$P_+(A_i)$	$P_-(A_i)$	$I_+(A_i)$	$I_-(A_i)$	I
0.50	0.50	1.00	1.00	1.00
0.90	0.10	0.15	3.32	0.47
0.99	0.01	0.013	6.64	0.078

extent (e.g., the presence of chromium in steel); $P_+(A_i) = 0.99$: the analyte is expected in the sample with high probability like manganese in steel. As can be seen at the right-hand side of the table, the average information content reaches its maximum if the a priori probabilities are equal. As I decreases, the more probable is the presence of the analyte. On the other hand, as the specific information content increases, the more improbable is the corresponding result.

In general, the uncertainty after analysis corresponds to some conditional probability (c.p.). We list these conditions as follows ($i = 1, 2, \ldots, n$; $j = 1, 2, \ldots, m$):

$P(z_j|A_i)$ c.p. that the signal occurs at position z_j if analyte A_i is present.

$P(z_j|x_i)$ c.p. that the signal at position z_j appears if the ith analyte is present in concentration x_i.

$P(y = 1|x_i)$ c.p. of the appearance of a signal whose position cannot be considered (e.g., a spot test[1] in chemical analysis) if the ith analyte is present in a concentration higher than the detection limit $x_i > x_{DL}$.

$P(y = 0|x_i) = 1 - P(y = 1|x_i)$ c.p. that a signal, as characterized above,[1] does not appear if the concentration of the ith analyte is $x_i < x_{DL}$.

$P(y = 1|A_i)$ c.p. for $P(y = 1|x_i)$ where x_i is much greater than the corresponding detection limit.

$P(A_i|y = 1)$ c.p. that the analyte A_i is present if the signal, as characterized above, occurs.

$P(A_i|z_j)$ c.p. that the analyte A_i is present if a signal at position z_j occurs.

$P(x_i|z_j)$ c.p. that the ith analyte is present in concentration x_i if the signal at position z_j appears.

$P(x_i|y = 1)$ c.p. that the ith analyte is present in a concentration $x_i > x_{DL}$ if a signal appears in chemical analysis as characterized in case of $P(y = 1|x_i)$.

---

[1]$y = 1$ means a visible trace of a chemical reaction (e.g., precipitate or color-change after a reagent is added); $y = 0$ means no change.

The probabilities $P(z_j|A_i)$, $P(z_j|x_i)$, $P(y = 1|x_i)$, $P(y = 0|x_i)$, and $P(y = 1|A_i)$ can be determined experimentally, since, for example, the frequency of appearance of a signal depends on the kind of analyte or on its concentration. The probabilities $P(A_i|y = 1)$, $P(A_i|z_j)$, $P(x_i|z_j)$, and $P(x_i|y = 1)$ are necessary in the signal decoding. They can be determined by BAYES' relation (3.6) for $P(y = 1|A_i)$, $P(z_j|A_i)$, $P(y = 1|x_i)$, or $P(z_j|x_i)$.

In instrumental analysis the signals are assigned to the components in a somewhat reliable way (Fig. 5.2). Under unfavorable circumstances the signals may overlap each other or appear distorted. In situations where analytes are detected at concentrations approaching the detection limit, the information content can be obtained by using the entropy for conditional probabilities; or the information gain can be calculated from the a priori conditional probabilities by using the divergence measure.

The uncertainty of a signal detection can be expressed by single-component information. For chemical detection the single-component expression has the form

$$I = \text{lb } 2 - \left(- \sum_{j=1}^{2} p_j \text{ lb } p_j \right) = 1 + \sum_{j=1}^{2} p_j \text{ lb } p_j \qquad (5.4)$$

where $p_1 = P(A_i|y = 1)$ and $p_2 = 1 - p_1$. In instrumental analysis, we would use $p_1 = P(A_i|z_j)$. Although Eq. (5.4) gives a simplification of the problem, the dependence of I on $P(A_i|z_j)$ is very instructive, and therefore some values are given in Table 5.2. The data in the table indicate that even a small uncertainty in the detection is associated with a large decrease in information content; for $p_1 = p_2 = 0.5$, where one cannot decide whether the analyte sought is present or not, the information content is zero. According to Eq. (4.11) component entropy is related to the selectivity criterion of detection. This criterion is more important in qualitative multicomponent analysis.

When several components are to be detected simultaneously or sequentially, the information amount M obtained from such multicomponent analysis is given by Eq. (3.2), where $m = \Sigma I_i$. Each of the n components to be analyzed is characterized by an a priori

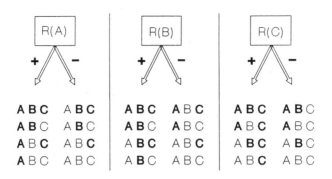

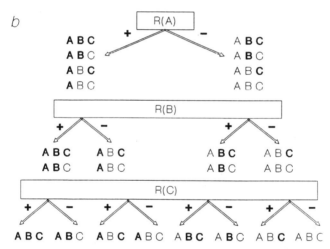

Figure 5.2. Qualitative analysis of three components A, B, and C by independent detection (*a*) and separation scheme (*b*). Components found to be present appear in bold type.

Table 5.2. Information Content I (in Bits) Dependent on the Conditional Probability $P(A_i|z_j)$

$P(A_i|z_j)$	1.00	0.99	0.95	0.90	0.80	0.70	0.60	≤ 0.50
I	1.00	0.92	0.71	0.53	0.28	0.12	0.03	0.00

uncertainty $P_+(A_i) = P_-(A_i) = 0.5$. Then, combining Eqs. (3.2) and (5.1), the information amount is expressed as

$$M = n \operatorname{lb} 2 = n \operatorname{bit} \tag{5.5}$$

The information amount is independent of the way in which the information is obtained. We will illustrate this by a comparison of the two analytical procedures in detecting three components A, B, and C: First, each component is analyzed independently by a specific reagent: R(A), R(B), R(C), for example, by spot test where the sequence of tests is optional. Next, the components are analyzed in a given sequence, for example, by a systematic separation scheme.

The analytical procedures and the information schemes of these two cases are shown in Fig. 5.2. In each case there exist k possible combinations for the presence and absence of A, B, and C. The number k results from combinatorial theory as

$$k = \sum_{k=0}^{n} c_k^n = \sum_{k=0}^{n} \binom{n}{k} = 2^n \tag{5.6}$$

For the given problem we have $n = 3$, and therefore $k = 2^n = 8$ possible combinations from which the existing combination will be determined. As can be seen from Fig. 5.2, both procedures will lead to a correct combination of analytes despite the different approaches.

Suppose that the sample we have to analyze contains components A and C. By procedure (1) we find the outcome of reaction R(A) to belong to the left-hand column, R(B) to belong to the corresponding right-hand column, and R(C) to belong to the left-hand column. The only adequate result in the columns is **ABC**. By procedure 2 we find a sequence of the reactions R(A), R(B), and R(C) with a direct result **ABC**. In both cases we need three elementary yes–no decisions to obtain the right result in accordance with Eqs. (5.5) and (5.6).

We need to point out some important distinctions between situations involving identification analysis and qualitative analysis, given three analytes. For the identification case concerning one out of three possible analytes, the information content is $I = \text{ld}\, 3 = 1.58$ bit in the simplest situation. For the identification of all three components, the information amount required is $M_{id} = \text{ld}\, 3 + \text{ld}\, 2(+\text{ld}\, 1) = 2.58$ bit. But for the qualitative analysis of three components it is $M_{ql} = 3\,\text{ld}\, 2 = 3$ bit.

When the a priori uncertainties are different, the information content for each component is calculated using Eq. (5.2):

$$M = -\sum_{i=1}^{n} \left[P_+(A_i)\text{lb}\, P_+(A_i) + P_-(A_i)\text{lb}\, P_-(A_i) \right]_i \quad (5.7)$$

If the detection is carried out by an instrumental method, we further apply Eq. (5.4) to get

$$M = \sum_{i=1}^{n} \left[\text{lb}\, 2 - (-p_+\, \text{lb}\, p_+) - p_-\, \text{lb}\, p_- \right]_i \quad (5.8)$$

In cases where there is an abundance of signals some instrumental methods, such as atomic spectroscopy, cannot assign all the signals to analytes in an unambiguous way. Therefore analysts frequently make use of the opportunity to determine the analyte by several signals belonging to the same component. Uncertainties are reduced signal after signal (see Table 5.2) as the probabilities for all signals are computed. The information gain shown in Table 5.3 might be obtained for one component whose several signals are inspected in a qualitative analysis. Suppose we want to determine the presence of a given element by means of atomic spectroscopy. Since we have no prior information, the a priori probability is

Table 5.3. Increase of Information Gain (in Bits) by Evaluation of Additional Signals (Given for Different Input and Output Probabilities).

A priori Probability (Input Certainty), $P_0(A_i)$	A posteriori Probability (Output Certainty), $P(A_i|z_j)$				
	0.900	0.950	0.990	0.999	1.000
0.50	0.53	0.72	0.92	0.99	1.00
0.90		0.19	0.39	0.46	0.47
0.95			0.20	0.27	0.28
0.99				0.07	0.08

$P_0(A_i) = 0.5$. After inspection of the first line we are only about 90% certain about its presence, so the information gain of this first step is 0.53 bit. Then we inspect a second line, and we are 99% certain that the element is present. The information gain of the second step is 0.39 bit and on the whole 0.92 bit. The last line in Table 5.3 shows that we can increase the certainty to 100% if we consider a third signal.

5.3   PRACTICAL APPLICATIONS

In practice whether an analytical problem is treated as an identification analysis or a qualitative analysis depends on the analyst's point of view. Some of the examples given in Section 4.3 can be handled from a qualitative perspective. The difference between qualitative and semi-quantitative analyses can also be blurred, since most qualitative tests permit (without incurring additional costs) major-, minor-, and trace constituents to be differentiated and compared among different samples, with results stated in terms of "more or less than."

Clearly, the information gain by such semi-quantitative results is higher than that of simple yes–no answers. After all, the differentiation is not just between two steps but between several: for example, (1) absence, (2) trace-, and (3) minor-, and (4) major constituent. If the a priori probabilities are equal, the information contents results in $P_0(1) = P_0(2) = P_0(3) = P_0(4) = 0.25$, to $I = 4$ $(-0.25 \, \text{lb} \, 0.25) = 2 \, \text{bit}$, which gives twice as much information as a simple decision on presence or absence.

The detection of key components leads to an identification of the attributes of a given material. The investigation of the genuineness and value of a yellow sheet metal given in Section 4.3. may also be treated by qualitative analysis of the two key elements Cu and Au. Had we a case of complete uncertainty, $P_0(\text{Cu}) = P_0(\text{Au}) = 0.5$, we would have obtained the information amount

$$M = I(\text{Cu}) + I(\text{Au}) = 2[2(-0.5 \, \text{ld} \, 0.5)] = 2 \, \text{bit}$$

If we assume the same a priori probabilities as given in Section 4.3, $P_0(\text{Cu}) = 0.75$ and $P_0(\text{Au}) = 0.20$, the information amount of the experiment would be

$$M = -(0.75 \operatorname{ld} 0.75 + 0.25 \operatorname{ld} 0.25) - (0.20 \operatorname{ld} 0.20 + 0.80 \operatorname{ld} 0.80)$$
$$= 1.53 \text{ bit}$$

The specific information gain depends on the actual result. If we found gold to be a major component and only traces of copper, than the specific information content of this result would be, according to Eq. (4.12),

$$I_{sp} = -\operatorname{ld} 0.20 = 2.32 \text{ bit}$$

and otherwise if copper were found

$$I_{sp} = -\operatorname{ld} 0.75 = 0.42 \text{ bit}$$

These specific information gains are in accordance with that of identification analysis.

In analytical practice sorting, discrimination, and identification of materials, especially of metals, are carried out by qualitative methods of analysis of one or some elements. Key elements in alloys can be analyzed rapidly and reliably by spot tests—and sometimes directly on the sample—hand spectroscopes ("metascopes"), and portable spectrometers.

Routine investigations of environmental subdivisions are carried out frequently by screening methods with chemical or instrumental procedures. Test sets containing test papers, reagents, and cuvettes for rapid colorimetric or photometric tests on organic and inorganic components are widely used for specific color reactions, as are portable X-ray spectrometers for the qualitative and semi-quantitative analysis of inorganic constituents.

Preliminary information about the subject under investigation is important for the information content of the qualitative analyses and the information gain of the results. In environmental analysis the detection limits are frequently stipulated by international limits and anything exceeding them has to be controlled.

While quantitative results can be evaluated and compared by statistical methods, qualitative results cannot. Information-theoretic quantities represent the mathematical criteria for the assessment and comparison of identification and qualitative analysis. The advantages of information-theoretic evaluation are particularly evident in multicomponent analyses.

About the different measures of information, it can be stated that, on the one had, the specific information content I_{sp} characterizes a concrete result of analysis and therefore the information gain by a special analysis; while, on the other, the (average) information content of an experiment characterizes all the possible results that can be obtained, in principle, by a special analytical method and therefore by the performance of the method itself. The average information content can be 1 bit at best in a qualitative analysis, but it can be higher in identification analysis, depending on the number of components out of which the unknown is to be selected.

More discussion of the information properties of qualitative analysis can be found in publications by CLEIJ and DIJKSTRA (1979), DANZER (1973), ECKSCHLAGER and KRÁL (1984), and LITEANU and RICA (1979b). We will discuss selectivity problems in detail in Chapter 7 on multicomponent analysis.

The probabilities $P(x_i|y = 1)$ and $P(x_i|z_j)$ are not of much use on their own. What is important is their dependence on x_i, which is the underlying principle of frequentometric analysis described by Liteanu and Rica in their monograph (1979a). In Chapter 6 and 7 where we discuss trace analysis and multicomponent analysis, we will consider this dependence in some depth. The publications by ECKSCHLAGER and STEPÁNEK (1978, 1981) also explain this dependence.

REFERENCES

Cleij, P., and Dijkstra, A. 1979. *Fresenius Z. Anal. Chem.* **298**, 97.

Danzer, K. 1973. *Z. Chem.* **13**, 229.

Danzer K., Eckschlager, K., and Wienke, D. 1987. *Fresenius Z. Anal. Chem.* **327**, 312.

Danzer, K., Eckschlager, K., and Matherny, M. 1989. *Fresenius Z. Anal. Chem.* **334**, 1.

Eckschlager, K. 1975. *Fresenius Z. Anal. Chem.* **227**, 1.

Eckschlager, K. 1991. *Collect. Czech. Chem. Commun.* **56**, 505.

Eckschlager, K., and Král, M. 1984. *Collect. Czech. Chem. Commun.* **49**, 2342.

Eckschlager, K., and Stepánek, V. 1978. *Mikrochim. Acta* **I**, 107.

Eckschlager, K., and Stepánek, V. 1979. *Information Theory as Applied to Chemical Analysis.* Wiley, New York.

Eckschlager, K., and Stepánek, V. 1981. *Mikrochim. Acta* **II**, 143.

Eckschlager, K., and Stepánek, V. 1985. *Analytical Measurement and Information. Advances in the Information Theoretic Approach to Chemical Analysis.* Research Studies Press, Letchworth.

Eckschlager, K., Stepánek, V., and Danzer, K. 1990. *J. Chemometrics* **4**, 1915.

Liteanu, C., and Rica, I. 1979a. *Statistical Theory and Methodology of Trace Analysis.* E. Horwood, Chichester.

Liteanu, C., and Rica, I. 1979b. *Anal. Chem.* **51**, 1986.

CHAPTER

6

QUANTITATIVE ANALYSIS

A quantitative analysis seeks to answer the question "How much?" That is, in which concentration, amount, or content is the analyte of interest present? With today's instruments, an analysis is able simultaneously to answer questions of "What" and "How much" in that qualitative and quantitative analysis is carried out simultaneously.

In classical analysis the procedure is designed for the specific analyte under consideration, and only the signal intensity, y, such as the volume of titrant taken up during a titration, is measured.

Figure 6.1 shows the parameters in instrumental analysis. This is a simplified representation of the four analytical parameters. In instrumental analysis the signal's intensity y is measured with respect to the signal's position z_i for the given analyte A_i; this is frequently stated in the form $y(z)$.

If the signal positions of the different analytes are sufficiently far apart, more than one analyte can be determined at a time, as will be shown in Chapter 7 where we discuss instrumental multicomponent analysis. However, multicomponent analysis cannot always be depended on when the determination of several components is involved; interactions among analytes and with other phenomena must be taken into account. Therefore, we will first deal with the information aspects of quantitative single-component analysis, although methods of multicomponent analysis are mostly employed in practice.

The process of gaining information about the samples being analyzed occurs in an analytical system that may be subdivided into sets of subsystems. One of the possibilities is shown in Fig. 6.2. Usually, the first (experimental) and second (evaluating) stages where the chemical or instrumental quantitative (single- or multicomponent) analysis takes place are not very complicated. Yet the system always consists of several subsystems.

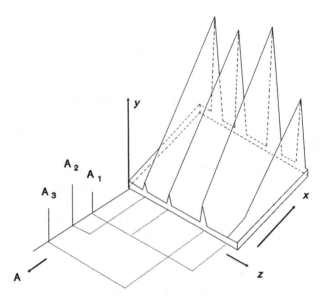

Figure 6.1. Simplified representation of the four-dimensional connection between different analytes A, the amount (concentration) of these analytes x, signal position z, and signal intensity y.

In the first stage, where information is created, the subsystems include sampling, sample decomposition and preparation, and occasionally separation and completion of a chemical reaction or a physical interaction. This stage is concluded by measurement. The second stage, namely evaluation, usually includes calibration and calculation of the analytical result and some precision characteristics such as the relative or absolute standard deviation and the confidence interval.

We imagine that the subsystem operations of the first and second stages proceed separately. Although the functions of the subsystems, their order and relatedness are necessary for the success of the whole process, each contributes a share to the random, and occasionally to the bias, component of the a posteriori uncertainty. In some cases the contributions can be found by calculations based on chemical and physical laws and rules, such as the chemical equilibrium law, as demonstrated by Eckschlager (1969, ch. 3).

The fact that the subsystem operations affect the final result is the main reason for seeking the optimum analytical stra-

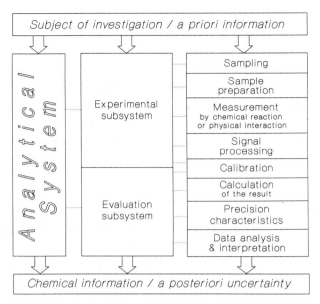

Figure 6.2. The analytical system and its stepwise subdivision into analytical subsystems.

tegy (Doerffel and Eckschlager 1981) and chemometric strategy (Doerffel, Eckschlager, and Henrion 1990): Both the operations and their interrelationships are optimized.

Also relevant here is the way in which information about signal intensity y or $y(z)$ is transformed into information about the chemical composition (*analytical information* into *chemical information*). To keep the quantitative analysis resistant to minor deviations from the optimal procedure ("robust"), the functioning of the quantitative-analytical system is continually checked by applying chemical, physical, and metrological provisions. The basis for that is given by validation principles and rules of quality assurance.

6.1   SINGLE-COMPONENT ANALYSIS

The decisive characteristic in quantitative analysis is the signal intensity y or $y(z)$, which is proportional to x, the amount, content, or concentration of the analyte being determined. The input–output relation is given by

$$y = f(x) + e_y \approx A_y x + e_y \qquad (6.1)$$

where $f(x)$ represents the underlying functional relation, e_y the measurement error for y, and A_y the (approximate) linear operator that transforms x into the expected value of y (Currie 1993).

In the case of two-dimensional analytical information we have:

$$y(z) = A(z)x + e_y(z) \qquad (6.2)$$

The input–output relation of a single-component analysis system is given by the stoichiometry of the analytical reaction in the case of chemical analysis, and by the empirically determined calibration function in the case of instrumental analysis. The analytical function is given by the inverse function of Eq. (6.1) or (6.2), respectively,

$$x = f^{-1}(y, e_y) \approx A_x(y, e_y) \qquad (6.3)$$

$$x = A^{-1}(z)[y(z), e(z)] \qquad (6.4)$$

For more details about calibration and analytical functions, see Eckschlager and Stepánek (1985, pp. 10–12) and Danzer (1994).

Important in using the divergence measure concept to determine the information content or gain are the forms of the a priori and a posteriori distributions. In determining the single main component, the a posteriori distribution is invariably normal, Gaussian, $N(\mu, \sigma^2)$, whose probability density follows Eq. (3.21). The most probable value of the result μ represents the required information, and the parameter σ^2 its precision. Since the earliest work on this topic (Eckschlager 1975) the a posteriori distribution was regarded as normal, and the divergence measure was employed to evaluate the information content of the quantitative result.

The a priori distribution, which reflects our assumption or preliminary information about the analytes in sample, is considered to be uniform, $U(x_1, x_2)$, by Eq. (3.18). Then if we know nothing about the analyte content, we can take values from x_1 to x_2 (e.g., from 0 to 100%) with the same probability. If, however, we have experimentally obtained preliminary information, we can consider the a priori distribution to be normal, $N(\mu_0, \sigma_0^2)$.

For a uniform a priori distribution and a normal a posteriori distribution—assuming that the results are accurate and that the expected μ value of the posteriori distribution lies within the limits of $x_1 + 3\sigma \leq \mu \leq x_2 - 3\sigma$, the information gain can be expressed

$$I(p, p_0) = \text{lb}\,\frac{x_2 - x_1}{\sigma\sqrt{(2\pi e)}} \tag{6.5}$$

In this relation x_1 and x_2 are the lower and upper limits, respectively, of the region within which the analyte content is expected prior to analysis, and σ characterizes the precision of the result; $\sqrt{(2\pi e)} = 4.1327$.

If the condition that $x_1 + 3\sigma \leq \mu \leq x_2 - 3\sigma$ is not met, or if $(x_2 - x_1) < 6\sigma$, the information gain is

$$I(p, p_0) = \text{lb}\,\frac{x_2 - x_1}{\sigma\sqrt{(2\pi e)}} - \text{lb}[\phi(z_2) - \phi(z_1)]$$

$$+ \frac{1}{2\ln 2}\,\frac{z_2\varphi(z_2) - z_1\varphi(z_1)}{\phi(z_2) - \phi(z_1)} \tag{6.6}$$

where $\varphi(z_i)$ and $\phi(z_i)$ are the frequency and distribution functions, respectively, of the normal distribution for the normalized random quantity $z_i = (x_i - \mu)/\sigma$, $i = 1, 2$.

If $\phi(z_2) - \phi(z_1) \geq 0.997$, Eq. (6.6) transforms into Eq. (6.5). Equation (6.6) characterizes the information gain of an incorrect a priori assumption, where the condition $\mu \in \langle x_1, x_2 \rangle$ is not fulfilled, or the relatively rare case of a uniform a priori distribution that is so narrow that $(x_2 - x_1) \leq 6\sigma$. In plain language it means that our expectation range cannot be reduced by the result and its confidence interval. For example; given an expected value in the interval between $x_1 = 16.5\%$ and $x_2 = 18.0\%$, the result is $\mu = (17.3 \pm 1.1)\%$. The relationship of the expectation range (rectangular area) to the a posteriori distribution can be seen in Fig. 6.3b.

Invariably the information gain obtained by Eqs. (6.5) and (6.6) is $I(p, p_0) > 0$ provided that the results are accurate. Figure 6.3

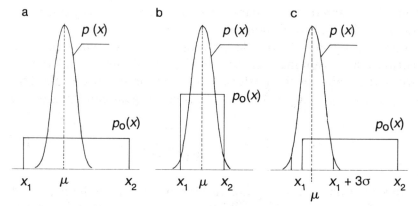

Figure 6.3. Different relations of the a priori and a posteriori distributions $p_0(x)$ and $p(x)$: (a) Eq. (6.5); (b), Eq. (6.6), $(x_2 - x_1) < 6\sigma$; (c) Eq. (6.6), $x_1 + 3\sigma > \mu$

compares the reduction of uncertainty between the a priori distribution $p_0(x)$ and the a posteriori distribution $p(x)$ for the general case given by Eq. (6.5).

In some cases it is difficult to set the limits of an a priori uncertainty in a realistic way. Then one could conceivably use a range $\langle 0, 100\% \rangle$ for general unknown cases, $\langle 0, 100 \text{ ppm} \rangle$ for ordinary trace analysis, $\langle 0, 1 \text{ ppm} \rangle$ for ultra-trace analysis, and so on. But what about routine analyses of well-known materials, proficiency testing, and round-robin tests? The opinions of analysts differ on this matter. We might ask skilled analysts in any field for the limits of an expectation range, and even after analysis when the information gain is calculated, they may be at a loss to specify the limits or may suggest limits that are too narrow.

It is clearly advantageous to establish general limits of the expectation range (the a priori uncertainty) for cases in which they are not obvious. Suppose we define

$$x_1 = \mu - \frac{\mu}{2} \approx \bar{x} - \frac{\bar{x}}{2}$$

$$x_2 = \mu + \frac{\mu}{2} \approx \bar{x} + \frac{\bar{x}}{2} \tag{6.7}$$

We have

$$(x_2 - x_1) = \mu \approx \bar{x} \tag{6.8}$$

If we introduce $m = (x_2 - x_1)/[\sigma\sqrt{(2\pi e)}]$ as the number of distinguishable concentration steps Doerffel and Hildebrandt 1969; Eckschlager 1971: Danzer 1973), Eq. (6.5) becomes

$$I(p, p_0) = \text{lb } m \tag{6.9}$$

Equations (6.5) and (6.9) can also be employed to express the information content of the analytical signal in a single-component analysis:

$$I_y = \text{lb} \frac{y_{\max} - y_{\min}}{\sigma_y\sqrt{(2\pi e)}} = \text{lb } m_y \tag{6.10}$$

Here $y_{\min}$ and $y_{\max}$ are the minimum and maximum signal intensities recorded by the instrument, and σ_y is the standard deviation for the signal measurement. These are all instrumental parameters. By Eq. (6.8), with $\sigma_y\sqrt{(2\pi e)} \approx 2s_y t_{a,f}/\sqrt{n} = 2\Delta\bar{y}$ where $\Delta\bar{y}$ is the confidence interval and $t_{a,f}$ a quantile of Student's t-distribution, we have for $y_{\max} \to \bar{y}$ and $y_{\min} \to 0$: $m_y = \bar{y}/(2\Delta\bar{y}) \approx S/N$, which corresponds to the signal-to-noise ratio (S/N) by which the information content is evaluated in information science and communication engineering.

Since the signal-to-noise ratio is occasionally also used in analytical chemistry, in particular with respect to the performance capability of instruments, some remarks about its usefulness in this connection should be made. But, first, we need to distinguish between the ordinary signal-to-noise ratio, S/N which is written

$$\frac{S}{N} = \frac{\bar{y}}{s_y} = \frac{1}{s_r} \tag{6.11}$$

(s_r relative standard deviation), and the signal-to-noise power ratio,

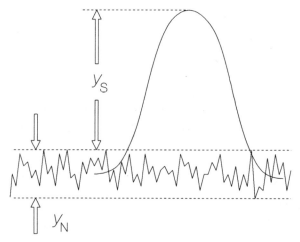

Figure 6.4. Parameters of the signal-to-noise ratio

P_S/P_N, which is written

$$\frac{P_S}{P_N} = \frac{y_S^2}{y_N^2} \tag{6.12}$$

where $P_S \sim y_S^2$ represents the signal power and $p_N \sim y_N^2$ the noise power (see Woschni 1970, 1972; Ziessow 1973; Danzer, Eckschlager, and Wienke 1987).

A visual explanation of the parameters in Eq. (6.12) is given in Fig. 6.4. The statistical risk of error $a = 0.01$ is the basis for the confidence interval of noise, which is written

$$s_N \approx s_y \approx \frac{y_N}{5} \tag{6.13}$$

The signal-to-noise ratio given by Eq. (6.11) is the criterion used to decide the precision of analytical instruments and procedures. It is identical with Pearson's coefficient of variability proposed by Kaiser and Specker (1956) who designated it as "precision" Γ for the characterization of analytical procedures.

Today the signal-to-noise ratio is important in analytical chemistry as automation and computerization proceeds. Fujimori,

Miyazu, and Ishikawa (1974) proposed its use for analytical balances and gravimetric analysis, still it has not been generally accepted for typical chemical analysis.

The signal-to-noise ratio and related instrumental parameters suggest that the information content of the signal conforms with the technical standard and quality of the instrument. This, however, is not entirely true for the information gain of the result. Information gain depends not only on the information content of the signal but also on the way the signal is processed, on preliminary information about the sample being analyzed, and on conditions in which the analysis is carried out.

If the preliminary information is the result of, say, a screening analysis with a normal distribution, then both the a priori and a posteriori distributions are normal. Information gain for the a priori and a posteriori distributions $N(\mu_0, \sigma_0^2)$ and $N(\mu, \sigma^2)$, with $\mu_0 = \mu$ and $\sigma_0 \geq \sigma$, respectively, is

$$I = \text{lb} \frac{\sigma_0}{\sigma} = -\frac{1}{2} \text{lb } R \qquad (6.14)$$

where $R - (\sigma/\sigma_0)^2$ is the reduction of variance, $R \in \langle 0, 1 \rangle$. This way of expressing information content or gain using the variance reduction concept was introduced by Kateman (1986); it provides a reasonable explanation of the nature of empirically obtained information and of the difference between information content (e.g., a signal) and information gain, or contribution, obtained by processing the signal into the result.

6.2   ACCURACY OF ANALYTICAL RESULTS

According to Eq. (6.14), information content depends only on variance reduction provided that good laboratory and analytical practice (GLP and GAP, respectively), and quality assurance of data prevent the appearance of a bias. However, information gain is not determined by variance reduction alone. Equation (6.5) can be written

$$I(p, p_0) = \mathrm{lb}\, \frac{\sigma_0 \sqrt{12}}{\sigma \sqrt{(2\pi e)}} = -\frac{1}{2} \mathrm{lb}\, R - 0.2546 \qquad (6.15)$$

since $\sigma_0^2 = (x_2 - x_1)^2/12$; however, it only holds true for $R \le 1/3$, which corresponds to $(x_2 - x_1)/\sigma \ge 6$. The term $\frac{1}{2} \mathrm{lb}\, [12/(2\pi e)] = -0.2546$ represents the difference, the divergence in the shapes of the uniform a priori and normal a posteriori distributions (see Fig. 6.3). But the information gain does not depend on the variance reduction alone even if the distributions are the same, normal. For instance, using the divergence measure, with $\mu \ne \mu_0$ and $\sigma \ne \sigma_0$, we have

$$I(p, p_0) = \mathrm{lb}\, \frac{\sigma_0}{\sigma} + \kappa \left[\left(\frac{\mu - \mu_0}{\sigma_0} \right)^2 + \frac{\sigma^2 - \sigma_0^2}{\sigma_0^2} \right] \qquad (6.16)$$

with $\kappa = 0.7214 = \frac{1}{2} \mathrm{lb}\, e = 1/(2.\ln 2)$. For the frequent case where $\sigma_0 > \sigma$, this relation, after allowing for variance reduction, can be written as

$$I(p, p_0) = -\frac{1}{2}\mathrm{lb}\, R + \kappa \left[D_0^2 + (R - 1) \right] \qquad (6.17)$$

The term

$$D_0^2 = \left(\frac{\mu - \mu_0}{\sigma_0} \right)^2 \ge 0 \qquad (6.18)$$

is not a metrological quantity (i.e., a quantity with analytical properties); rather, it might be called an "element of surprise" in the result. If μ_0 is the a priori assumption (following from theory, experience, or preliminary information), then D_0^2 will be larger the more the obtained result μ differs from μ_0. As μ approaches to μ_0, it follows that $D_0 \to 0$. Therefore, to characterize the *plausibility* of this occurring, we can write

$$P_D = \frac{1}{D_0^{\,2}} \qquad (6.19)$$

P_D will be the larger the more the result μ agrees with the theoretical assumption μ_0.

Information content I, expressed as the difference between the a priori and a posteriori entropy, is different in meaning from the information gain of accurate results $I(p, p_0)$ expressed by the divergence measure (i.e., as the difference between the Kerridge's and Bongard's measures for the a priori and the a posteriori distributions); for $\delta = 0$, $H(p, p)$ transforms into the entropy $H(p)$, see Table 3.3.

Information content generally concerns the analytical signal; hence it represents an intermediate but important step on the way to analytical information about a material sample. Information gain refers to the final result. The information content gives some facts about the properties of the signal after they are decoded, whether or not they are of interest to us.

Let us consider comparison of an a priori rectangular or normal distribution with an a posteriori normal distribution, which is facilitated by Figs. 6.3 and 6.5. In the case of the a priori uniform distribution, where $\mu \in \langle x_1, x_2 \rangle$, the information content I attains

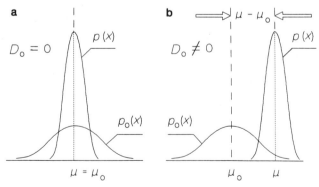

Figure 6.5. Divergence of a priori and a posteriori normal distributions given by $\sigma \neq \sigma_0$ (a) and $\mu \neq \mu_0$ and $\sigma \neq \sigma_0$ (b)

the same value as the information gain $I(p, p_0)$ obtained by Eq. (6.5) in that it only depends on the variance reduction, and the constant value of lb $\sqrt{[12/(2\pi e)]} = -0.2546$ represents the divergence in shape of the a priori and a posteriori distributions. In Eq. (6.5) $I(p, p_0)$ contains no element of surprise because we a priori expect any $\mu \in \langle x_1, x_2 \rangle$ with the same probability. For both the a priori and a posteriori normal distributions, the information content is determined by the variance reduction as in Eq. (6.14) where the information gain $I(p, p_0)$ contains, in addition, the term D_0^2 which characterizes the plausibility of the result.

Equation (6.16) does not transform into Eq. (6.14) even for $\mu = \mu_0$; only if the both distributions (a priori and a posteriori) are the same, can we have $I(p, p_0) = I = 0$. (The relation (6.16), as a model of information gain for higher-precision analysis is discussed in detail in the monograph by Eckschlager and Stepánek 1979, ch. 6.2.) As is clear from Eq. (6.5) the dependence of $I(p, p_0)$ on $-\log R$ for $R < \frac{1}{3}$ is linear, as $I(p, p_0)$ depends by Eqs. (6.14) and (6.16) or Eq. (6.17), on $-\log R$ (see Section 6.6., Fig. 6.20). Invariably, however, $I(p, p_0) \geq 0$.

The results of quantitative analysis should not but may involve a bias

$$\delta = X - \mu \qquad (6.20)$$

where X and μ are the true and observed analyte amounts respectively. Information gain here must be expressed by an extended divergence measure as the difference between Kerridge's and Bongard's measures for the a priori or a posteriori distributions and for the true distribution $r(x)$. The information gain for the a priori uniform distribution $p_0(x) \to U(x_1, x_2)$, the a posteriori normal distribution $p(x) \to N(\mu, \sigma^2)$, and the hypothetical true normal distribution $r(x) \to N(X, \sigma_r)$ (i.e., $\mu_r = X$), as represented in Fig. 6.6 is

$$I(r; p, p_0) = \text{lb} \frac{x_2 - x_1}{\sigma \sqrt{(2\pi e^k)}} - \frac{\text{lb } e}{2} \left(\frac{\delta}{\sigma} \right)^2 \qquad (6.21)$$

where $k = (\sigma_r / \sigma)^2$ [for a derivation of Eq. (6.21), see Eckschlager,

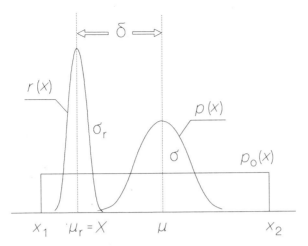

Figure 6.6. Comparison of a priori uniform, a posteriori normal and true value normal distributions

Stepánek, and Danzer 1990]. Since k is known, we have

$$I(r; p, p_0) = \text{lb}\frac{x_2 - x_1}{\sigma\sqrt{(2\pi)}} - \kappa\left[\left(\frac{\delta}{\sigma}\right)^2 + k\right]$$

$$= \text{lb}\frac{x_2 - x_1}{\sigma\sqrt{(2\pi)}} - \kappa\left(\frac{\delta^2 + \sigma_r^2}{\sigma^2}\right) \qquad (6.22)$$

with $\kappa = \frac{1}{2}$ lb e $= 0.7214$.

Since k for given σ_r depends on σ, the information gain in Eqs. (6.21) and (6.22) for $\delta = 0$ differs from that in Eq. (6.5) or (6.6) where rather the information content depends on σ. The relation is shown in Fig. 6.7.

An error δ is factored into chemical analytical procedure involving "nonquantitative" chemical equilibrium; in instrumental procedures it accounts for imprecise calibration. Precise calibration assures us that no quantitative deviation is present (e.g., as in optical emission spectrography).

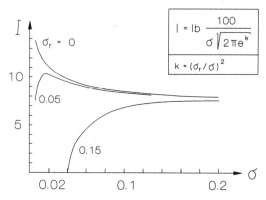

Figure 6.7. The difference between information content and gain depending on σ

Introducing the variance reduction $R = (\sigma/\sigma_0)^2$ into Eqs. (6.21) and (6.22), we find the information contribution:

$$I(r; p, p_0) = -\frac{1}{2}\operatorname{lb} R + \frac{1}{2}\operatorname{lb}\frac{12}{2\pi e^k} - \kappa\left(\frac{\delta}{\sigma}\right)^2$$

$$= -\frac{1}{2}\operatorname{lb} R + \frac{1}{2}\operatorname{lb}\frac{12}{2\pi e} - \kappa\left[\left(\frac{\delta}{\sigma}\right)^2 + k\right]$$

$$= -\frac{1}{2}\operatorname{lb} R + \frac{1}{2}\operatorname{lb}\frac{12}{2\pi e} - \kappa\left(\frac{\delta^2 + \sigma_r^2}{\sigma^2}\right) \quad (6.23)$$

For the a priori normal distribution $p_0(x) \to N(\mu_0, \sigma_0^2)$, a posteriori normal distribution $p(x) \to N(\mu, \sigma^2)$, and true normal distribution $r(x) \to N(X, \sigma_r^2)$, the information content is

$$I(r; p, p_0) = \operatorname{lb}\frac{\sigma_0}{\sigma} + \kappa\left[\left(\frac{\mu - \mu_0}{\sigma_0}\right)^2 + k\left(\frac{\sigma^2 - \sigma_0^2}{\sigma_0^2}\right)\right] - \kappa\left(\frac{\delta}{\sigma}\right)^2$$

$$(6.24)$$

With variance reduction and $X \approx \mu$, it becomes

$$I(r; p, p_0) = -\frac{1}{2} \text{ lb } R + \kappa \left[D_0^2 + k(R - 1) \right] - \kappa \left(\frac{\delta}{\sigma} \right)^2 \quad (6.25)$$

which for a low k (i.e., for the proven analytical method) transforms into

$$I(r; p, p_0) \approx -\frac{1}{2} \text{ lb } R + \kappa \left[D_0^2 - \left(\frac{\delta}{\sigma} \right)^2 \right] \quad (6.26)$$

Whether or not the a priori distribution is uniform or normal, the information contribution increases with a decreasing variance reduction and with an increasing $k = (\sigma_r/\sigma)^2$ ratio. It becomes greater than zero (even if $R = 0$) when no variance reduction occurs, provided that the results are accurate ($\delta = 0$). The nonzero value is due to the divergence of the a priori and a posteriori distributions; it is not relevant whether shape divergence [the term lb $[12/2\pi e)]$ in Eq. (6.23)] or mere displacement [D_0^2 in Eq. (6.25)] is involved.

Now let us consider the true distribution $r(x)$ and the variance ratio $k = (\sigma_r/\sigma)^2 \geq 0$. We do not need to know the hypothetical normal distribution $r(x)$; it is sufficient that its expected value and variance are known. Since this distribution concerns exact results, we choose $r(x)$ to be the same as $p(x)$, which in a quantitative analysis of major components is the normal distribution with the expected value corresponding to the true result $\mu_r = X$; the variance can be simplified as $\sigma_r^2 = 0$, which may nevertheless be too much of a simplification since the variance value characterizes the precision with which we know X (i.e., the analyte content of the reference material by which we verify the accuracy of the analytical method in question).

The effect of $k = (\sigma_r/\sigma)^2$ will be demonstrated in a comparison of information gain for the case where we assume that the results are accurate and the case where we have verified experimentally that the results really are accurate. In the first case, where we assume that $\delta = 0$, the information contribution for the a priori rectangular distribution is calculated according to Eqs. (6.5) and (6.15). In the second case, after we verify experimentally that $\delta = 0$, we use Eq. (6.21) or (6.23) to find the information contribution. Comparing these equations, we have for $\delta = 0$,

$$I(p, p_0) = -\frac{1}{2} \text{lb } R + \frac{1}{2} \text{lb} \frac{12}{2\pi e}$$

$$= -\frac{1}{2} \text{lb } R + \frac{1}{2} \text{lb} \frac{12}{2\pi} - \kappa \qquad \text{(case 1)}$$

$$I(r; p, p_0) = -\frac{1}{2} \text{lb } R + \frac{1}{2} \text{lb} \frac{12}{2\pi e^k}$$

$$= -\frac{1}{2} \text{lb } R + \frac{1}{2} \text{lb} \frac{12}{2\pi} - \kappa k \qquad \text{(case 2)}$$

Since as a rule $R < 1$, we always have for $k < 1$, $I(r; p, p_0) > I(p, p_0)$; the difference $I(r; p, p_0) - I(p, p_0)$ can reach 0.7214 bit, which is no negligible value. For both normal distributions and for $\delta = 0$, we have

$$I(p, p_0) = -\tfrac{1}{2} \text{lb } R + \kappa\left[D_0{}^2 + (R - 1)\right] \qquad \text{(case 1)}$$

$$I(r; p, p_0) = -\tfrac{1}{2} \text{lb } R + \kappa\left[D_0{}^2 + k(R - 1)\right] \qquad \text{(case 2)}$$

Here we also have $I(r; p, p_0) > I(p, p_0)$ whose difference depends on variance reduction R.

6.3   EFFECT OF CALIBRATION ON INFORMATION GAIN

The information gain of quantitative analysis always depends on the kind and quality of calibration. The modelling of the relationship between signal intensity y and the amount, content or concentration x of analyte according to Eqs. (6.1) through (6.4) can take various forms.

Calibration is the best way to exclude systematic errors experimentally. Because of the reduction of bias calibration increases the information gain. But calibration requires additional measurements and therefore increases the chances of random error, and accord-

Table 6.1 Influence of experimental Calibration Methods on the Error of x Values

Calibration Dependence	Calibration Methods	Dependence of σ_x on σ_y
$y = bx$	Calibration of straight line	$\sigma_x = \sigma_y/b\sqrt{1/m + 1/n + \left[\sigma_b/(b\sigma_y)\right]^2 y^2}$
	Standard addition	$\sigma_x = \sigma_y/b(1 + 1/q)\sqrt{\left[2(1 - r_{xy})\right]}$
$y = a + bx$	Calibration of straight line	$\sigma_x = \sigma_y/b\sqrt{1/m + 1/n + \left[\sigma_b/(b\sigma_y)\right]^2 (y - \langle y \rangle)^2}$

Note: a, b regression coefficients, m number of points included in straight-line construction of the calibration, n number of replicate measurements in each point, $q = x_s/x$ ratio of standard addition to analyte concentration, r_{xy} correlation coefficient, $\langle y \rangle$ estimate of the mean of y values.

ingly error propagation. As a result the calibration error is as important a factor in experimental calibration as is the selection of the correct regression model (linear or nonlinear, weighted or unweighted, one or both variables error model, parametric or robust, etc.; see Danzer 1990).

In the simplest case the relation between the analyte amount and the signal intensity is functional; it is given by a direct proportionality $y = \beta x$, where β is a stoichiometric factor, (hence a constant known in advance). Since the precision of stoichiometric factors is usually high (atomic and molecular weight relations as a rule), the error of β can be neglected; the error of concentration values amounts to $\sigma_x = \sigma_y/\beta$.

In general, calibration is carried out experimentally. The most important procedures can be seen from Table 6.1. In each case the information gain increases with the sensitivity b (the slope of the straight-line calibration, the number of calibration points, and the number of replicates. The dependence of information gain on all these parameters is shown in more detail, by Danzer, Eckschlager, and Wienke 1987; the effect of the quality of reference materials on information gain resulting from an activation analysis, see Obrusnik and Eckschlager 1993.

6.4 APPLICATION OF REFERENCE MATERIALS
FOR QUALITY ASSESSMENT

The quality of a method and result is usually assessed by looking at
the reference material (RM). RMs can differ greatly in quality,
depending on the following:

1. The kind of material (pure metals or well-defined salts, compound environmental, clinical, and organ samples, etc.)
2. Origin of the material (natural or synthetic).
3. Characterization of the material (e.g., by interlaboratory comparisons; see Chapter 9).
4. Keeping quality (packing, storage, conservation etc.).

Depending on the material and the analytical problem, we can
sometimes assume that the stated value of a certified RM (CRM) is
almost error free and unbiased ($\sigma_r \rightarrow 0$). But more often we must
assume that reference values have errors and therefore we use
information-theoretic distribution models that allow us to consider
the imprecision in concentration values of RMs.

In our discussion here we will consider the work of Obrusnik and
Eckschlager (1993) who used an *instrumental neutron activation
analysis* (*INAA*) for assessing the quality of RMs. Even with INAA,
which can frequently produce fairly accurate results with low or
practically no bias, biased results cannot be entirely avoided. Therefore, for information-theoretic evaluation we would still use Eqs.
(6.21) and (6.22).

Since INAA involves the measurement of radioactivity as a last
step, theoretical values of σ, and often also δ, are calculated . Then
theoretical dependences of information values on σ, or δ, are
derived (see Obrusnik and Eckschlager 1993). The analytical procedures of Eqs. (6.21) and (6.22) make it possible to compare accurate
and inaccurate (biased) results for the value of information gain.
The dependence of information gain on σ for different values of
relative bias δ_r, as calculated by Eq. (6.21) with $x_2 - x_1 = 1000$
ppm, $x = 100$ ppm, and $k = 1$, is shown in Fig. 6.8.

In the figure it can be seen that the information gain for
unbiased results increases with decreasing σ even for very low

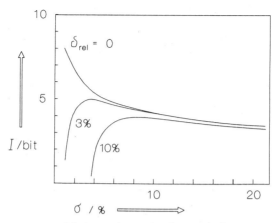

Figure 6.8. Dependence of information gain I on relative error σ and relative bias δ

values of σ (see the curve for $\delta = 0$). However, for biased results (3% or 10%), the curves of information gain decrease rather rapidly in the region of highly precise (low σ) results. The higher the level of bias, the lower is the value of the information gain obtained.

The results in Fig. 6.8 can be used to demonstrate the influence of *calibration procedures* on information gain with INAA. Calibration by synthetic standards prepared from pure elements and compounds can produce a relative bias of up to 3%. The information gain by INAA with this kind of calibration is depicted by the area between the I curves for the 0% and 3% biases. This information gain is fairly high. The δ value becomes statistically significant only in high precision cases, below $\sigma = 3\%$, and then information gain decreases. But, by using certified reference materials (CRMs) for calibration, with σ_r up to 20% for some elements, we can expect to obtain mostly low values of information gain. Clearly information theory confirms the disadvantage of using CRMs for calibration in INAA, pointed out by Becker (1987) and Heydorn (1988).

As mentioned earlier, the parameter $k = (\sigma_r/\sigma)^2$ indicates the reliability of a quality assessment procedure. Equation (6.21) for $\delta = 0$ presents a case where the existence of a nonzero bias is admitted, though it is proved experimentally that $\delta = 0$. Therefore for $k < 1$ the information gain according to Eq. (6.21) is higher,

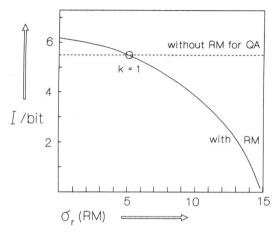

Figure 6.9. The effect of precision in the RM value on information gain; $\sigma = 5\%$, $\delta = 0$

then that found by Eq. (6.5) where the presence of a bias is not assumed. This difference results in a contribution to information gain since a quality assessment is used. The difference depends on the quality (reliability) of the RM used (σ_r), and it can be as high as $\kappa \approx 0.72$ bit; see Fig. 6.9 (the parameters chosen are the same as in Fig. 6.8).

The absolute values of σ and σ_r substituted, by way of k, into Eq. (6.21) or (6.22), respectively, are expressed (in concentration units) in both unknown and reference samples. Including the σ_r value might be a problem since not all producers of CRMs give this value on their certificates.

If precision is not an issue, it is possible to estimate σ_r by using an approximate number n for finding the certified value as $n \in \langle 15, 20 \rangle$, which is usually expressed as a 0.95 confidence interval (see Musil 1986). In general, it is permissible to assume that the matrix of RM is very similar to that of the sample.

The practical aspects of using information theory to choose the optimum RM for quality assessment is illustrated by Figs. 6.10 and 6.11. Fig 6.10 shows five CRMs that might be used to determine the arsenic in the fly ash matrix. These CRMs differ according to producer, in their certified concentrations, and in their uncertain-

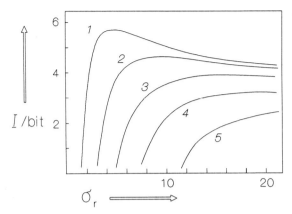

Figure 6.10. Use of CRMs for quality assessment in determination of As in fly ash

ties (and σ_r) of these concentrations as follows:

1: BCR CRM 038	Fly ash	48 ppm ± 4.8%
2: IRANT ECH	Fly ash	56.9 ppm ± 7.6%
3: IRANT EOP	Fly ash	79.1 ppm ± 8.4%
4: NIST SRM 1648	Urban particulate	115 ppm ± 8.7%
5: NIST SRM 1633a	Fly ash	145 ppm ± 10%

Since we can assume that the concentration range of As in analyzed samples of fly ash is a priori known to be less than 1000 ppm and real As concentrations in most of the samples are close to 100 ppm, we use $x_2 - x_1 = 1000$ ppm, $x = 100$ ppm, in Eq. (6.21) for the representation in Fig. 6.10.

The dependence of information gain on the value of σ for the above-mentioned CRMs has a maximum at $k = 1$ ($\sigma_r = \sigma$). It can be seen that under the given conditions, the BCR 038 fly ash gives the highest information gain when used for quality assessment purposes. Other CRMs have higher certified values, larger relative uncertainty ranges, and consequently higher σ_r values. The disadvantage of using SRM 1633a in this case is its relatively high σ_r value. But this CRM is adequate for samples containing between 150 and 200 ppm of As.

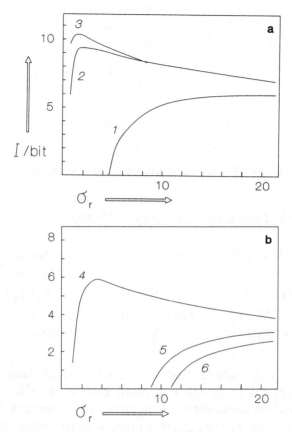

Figure 6.11. CRMs for quality assessment of Cd traces in biological material; (*a*) higher trace contents: $\delta = 0$, $x_2 - x_1 = 300$ ppm, $x = 5$ ppm; (*b*) low trace contents: $\delta = 0$, $x_2 - x_1 = 0.1$ ppm, $x = 0.01$ ppm

1: NIST SRM 1566	Oyster	3.5 ppm ± 11%
2: NIST SRM 1577a	Bovine Liver	0.44 ppm ± 14%
3: NIES CRM 6	Mussel	0.82 ppm ± 3.5%
4: NIST SRM 1549	Milk (powder)	0.0005 ppm ± 40%
5: BCR CRM 063	Milk	0.0029 ppm ± 41%
6: BCR CRM 150	Milk (spiked)	0.0218 ppm ± 6.4%

Determining the As content in fly ash is a relatively simple problem. Often situations are much more complicated, such as determining Cd traces in biological materials.

The possible Cd concentrations can range to almost six orders of magnitude, from 0.0002 to 200 ppm. Only the use of CRMs with concentrations similar or lower than the concentrations of Cd in real samples yields reasonably high information gain.

Figure 6.11 show two extreme cases of Cd concentrations: 5 and 0.01 ppm with expectation intervals of 300 and 0.1 ppm, respectively. For the higher concentration region in Fig. 6.11a NIES CRM 6 mussel and NIST SRM 1577a bovine liver give the best information gain while NIST SRM 1566 oyster tissue has a relatively high absolute σ_r value (as well as a higher concentration level and uncertainty). The use of IAEA H 8 horse kidney CRM with a certified Cd concentration of 189 ppm $\pm 2.4\%$ does not contribute any positive information, since the certified level (and σ_r) is too high compared with the Cd concentration in the analyzed samples.

Figure 6.11b gives a contrasting case where the Cd concentrations are at the 0.01 ppm level. The only CRMs that can bring positive information gain are CRMs 4–6. The NIST SRM 1549 milk powder is the best CRM here, even though the uncertainty of the certified concentrations is about 40%. An explanation for this is that σ_r (expressed in concentration units) is more important than the relative value as is its ratio to σ (also expressed in concentration units). For instance, these σ_r values are 0.00043, 0.0026, and 0.003 ppm for SRM 1549, CRM 063, and CRM 150, respectively. Other cases may be explained in the same way.

Figure 6.11 shows the wide range of dependence of information gain on σ values. Usually, we cannot expect to achieve better precision than between about 5% and 10%, and in extreme cases between about 10% and 20%. The steep decrease in information gain for more precise results (practically all the cases shown in Fig. 6.11) may be explained by the fact we cannot use CRMS with much higher σ_r levels to assess the bias of such precise results.

From these applications of information theory to quality assessment by means of CRMs, it should be clear the CRM chosen must be suitable for the concentration level of the analyte. While CRMs with substantially higher certified concentrations and low relative uncertainties can be used advantageously for the assessment of calibration, the absolute value (in concentration units) of the CRM's uncertainty has a crucial effect in the assessment of quality as was pointed out by Rasberry 1988).

Information theory proves that for extreme trace analyses even the use of reference materials with relatively wide uncertainty intervals of about 40% in the case of SRM 1549 milk powder can positively influence the information gain of the results. From this perspective the so-called second-generation CRMs for very low "natural levels" of elemental concentrations, such as human serum (J. Versieck et al. 1988), may prove to be very useful.

6.5   TRACE ANALYSIS

Trace analysis, whether qualitative or quantitative, does not differ much from the analysis of major components. However, since very low analyte contents are to be detected among high excesses of the matrix, some adverse features specific to trace analysis are effected, and this has led to the development of a separate branch of analytical chemistry. This analytical branch is also one of the areas of practical analytics receiving special attention.

In an information-theoretic evaluation the results and methods of trace analysis require a somewhat different approach from what was satisfactory for major component analysis. In particular, the following facts must be taken into account:

1. The a posteriori uncertainty is increased by effects such as sample contamination or loss of analyte (e.g., by evaporation of volatile components or by sorption on the vessel walls). While these effects occur during the analysis of higher contents as well, they are small enough to be disregarded. Some of these factors, such as contamination, do not lend themselves to description by a suitable mathematical model.

2. The a posteriori distribution largely cannot be regarded as normal: It is distorted due to the fact that the signal corresponding to the analyte concentration cannot be discerned from the background noise or the signal-to-noise ratio is low,

the background is not zero, and so on. All this must be taken into account when choosing a suitable distribution.

3. The standard deviation is usually both high and imprecise, so even a high bias appears as statistically insignificant. Some authors even suggest that the notation of the bias loses practical meaning to trace analysis. Irrespective of whether we agree with this opinion or not, it is evident that, just due to the low relative precision, the results of trace analysis require an approach different from that conventionally applied to the determination of higher contents.

4. Sensitivity plays an important role in trace analysis. Therefore calibration is a major contributor to the uncertainty of the results. While standards and reference materials are usually lacking, matrix effects are considerable

5. Unless the trace analyte determination is selective, the intense signals of the major components in a multicomponent analysis interfere with the weak signals of the trace analytes. Removal of the main fraction of the matrix, or conversely, enrichment of the trace components, are operations that often distort the results.

6. Valuable information can be obtained not only by detection of an analyte and determination of its amount but also by nondetection that affords a proof of absence of a certain component. This case is especially important for analysis of high-purity materials. For this purpose in some cases the guarantee of an absence of certain elements may bring a higher information gain than their detection and quantitative determination.

The system in which qualitative or quantitative trace analysis takes place always includes a high-sensitivity instrument. The measuring operation itself, however, has not such a dominant position as it has in the analytics of higher contents. In trace analysis major attention must be paid to sample preparation for measurement and to signal decoding, which is usually rather intricate and occasionally includes blank correction.

Data accumulation, filtering, signal smoothing, various transformations, and other procedures are sometimes applied for the enhancement of the signal-to-noise ratio or/and the improvement of the signal resolution to better finally the detection limit.

6.5.1 Methods of Trace Analysis

Trace analysis is one of the most important areas in analytical chemistry nowadays. Trace components play an essential role in materials science, environmental research, medical science and pharmacology, and microelectronics. In general, methods with high specifity [see Section 4.1, Eqs. (4.2) and (4.3)], detection power, precision, and accuracy are used for trace analyses. In most cases sensitive multielemental determinations must be conducted [Eq. (4.3)]. Additional marginal conditions may be nondestructive analysis, limited sample size, and so on. As a rule, not all such requirements can be fulfilled simultaneously, not only because of insufficiencies of instruments and equipments but also because of somewhat closed and complex connections among several parameters. One such connection concerns the regular increase of the random error with decreasing sample amount.

Another relation is described by the so-called system of analytical working ranges (IUPAC 1979) whereby the sample mass, $S =$

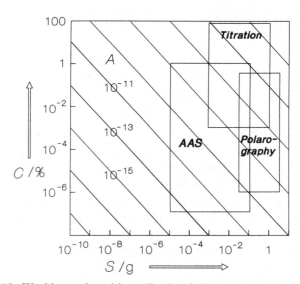

Figure 6.12. Working scales with application fields for the analysis of liquids

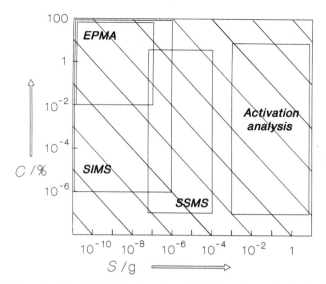

Figure 6.13. Application fields of selected methods for the analysis of solids within their analytical working ranges

$m_A + m_M$, the absolute analyte mass, $A = m_A$, and the analyte content,

$$C = \frac{m_A}{m_A + m_M} = \frac{A}{S} \qquad (6.27)$$

are connected with each other (m_M refers to the mass of the matrix, which consists of all the constituents of the sample except the analyte itself).

Figure 6.12 shows the application fields for selected methods of analysis for liquids, and Fig. 6.13 gives selected methods for the analysis of solids. As can be seen, the ranges of analyte mass and sample weight determine the range of analyte content, and therefore the detection limit is fixed.

Modern trace analysis can take one of two forms, as shown in Fig. 6.14:

1. A *direct* instrumental *method*.
2. Combined *multiple-stage methods*.

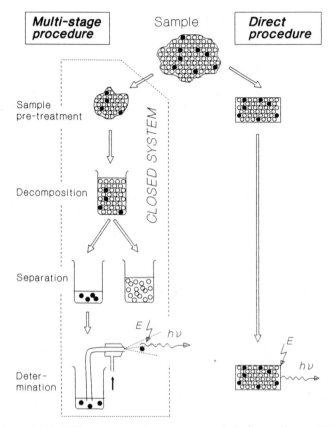

Figure 6.14. Different strategies in trace analysis (according to Tölg)

In a direct analysis, after an initial preparatory step involving physical setup, the sample is processed right up to the final result. The following methods are suitable for direct trace analysis:

- *Optical atomic emission spectroscopy*, especially by *arc* excitation, is widely used because of its high performance-to-cost relation. About 90% of the elements can by analyzed up to a lower ppm range, with relative uncertainty of 20% to 50%.
- *X-ray fluorescence spectrometry* (XFS, XFA) in its classical methodology is generally not a prominent trace method. Detection limits of elements depend on the atomic number of the

analyte as well as the matrix, and therefore only selected elements can be detected within the limits of a ppm level. The development of *Total reflection X-ray fluorescence analysis* (TXFA) has decisively changed the situation. With very low detection limits that can approach the ppb and ppt levels in some cases, TXFA is one of the most powerful trace analytical method nowadays. TXFA is a suitable method for the analysis of elements with atomic numbers larger than 11, especially heavy elements.

- *Solid state mass spectroscopy* (SSMS; the abbreviation stands also for *Spark source mass spectroscopy*, one of the most suitable techniques of inorganic mass spectroscopy for trace analysis) is the most powerful method of classic trace analysis because it can analyze practically all elements with high detection power up to the ppb level. Besides the spark source, laser ablation (LA) is used to transport sample material into the mass spectrometer (LA–MS), occasionally over the roundabout way of an inductively coupled plasma (LA–ICP–MS).

- *Secondary ion mass spectrometry* (SIMS) ranks high among the microanalytical techniques that are primarily characterized by low absolute detection limits. Whereas micro- and distribution-analytical methods generally have unfavorable relative detection limits, SIMS distinguishes itself by low determinable contents, namely relative detection limits.

- *Neutron activation analysis*, particularly in its instrumental variant (INNA), is surpassed in its detection power only by few methods. For many elements detection limits up to the ppb level can be attained, depending on the cross section and half-life of the activated analyte as well as the matrix. A particular advantage here may be that rare elements frequently have a larger cross section. Therefore such rare elements can be determined with low detection limits unlike most other methods.

Direct instrumental methods consist mostly of complex physical interactions not only with the analyte but also with other elements of the matrix (matrix effects). Only in exceptional cases will the matrix effects be treated theoretically and corrected by calculation

(XRF). In general, the matrix effects are considered by calibration, particularly in terms of reference materials, have a composition very similar to that of the sample under investigation. CRMs exist only for a few important substance groups and hardly for any ultra trace analyses. Although standard addition techniques may be considered, they are problematic and possible only in exceptional cases.

An alternative to the direct method is the so-called *multiple-stage procedure*—with coupled or combined techniques—see Fig. 6.14. The notion of a coupled procedure and its execution in a closed system was developed by Tölg and his coworkers (1979, 1987). All the operations required to produce a sample that can be measured by a selected determination procedure (e.g., a solution containing the analyte in an enriched form) take place one after the other without contact with the environment. There are two essential approaches:

1. Chemical decomposion by means of heating, microwaves, and UV radiation.
2. Separation and enrichment by chromatographic methods (GC, HPLC), ion exchange, liquid extraction, solid phase extraction (SPC), and electrolysis.

Some of these coupling procedures are synchronous with fixed instruments. Among these so-called *hyphenated techniques* are, in particular, GC–MS, HPLC–MS (thermo- and electrospray MS), GC–AAS, HPLC–ICP, partly combined with flow techniques (FIA–SPE–AAS).

Multiple-stage procedures are advantageous if they can be carried out in clean rooms or "super-clean" laboratories. When analyzing extremely low contents, a special kind of trace-analytical "hygiene" precaution must be taken to prevent contamination from solvents, reagents, vessels, equipment, and the environment (i.e., laboratory atmosphere and dust emission of laboratory staff).

6.5.2   Information Content and Information Gain in Trace Analysis

In trace analysis the signal position and intensity are complemented by another important circumstance, namely the discernibility of the signal intensity from background. If $y(z)$ is so low that the signal cannot be discerned from background, the information content of the "missing" signal must be considered quite differently from any signal intensity that can be measured with high or low precision. If the signal can be discerned from background but the signal-to-noise ratio is low, the analyte will merely be detected; if the signal-to-noise ration is high enough, quantitative determination might also be accomplished. Thus the difference between qualitative and quantitative trace analysis is given not only by whether we a priori ask "what" or "how much" but also by the response of the system to the material input: by the intensity of the signal at the given position $y(z)$. Yet, since noise increases in parallel with the signal, an increase in detector sensitivity usually does not result in a shift of the boundary between qualitative and quantitative analysis. In other words, the signal-to-noise ratio, which is decisive for the boundary, is little affected.

The relationship between the input and output of trace analysis is given by the calibration dependence, which often can be determined only with low precision. The calibration dependence is usually heteroscedastic—the relative standard deviation is usually constant (or nearly constant), even across concentration spans of several orders of magnitude, while the probability distribution of the signal or results is seldom normal.

The *limit of detection*, defined by Kaiser in 1965, is an important notion in information-theoretic considerations of trace analysis. The limit of detection is the concentration of analyte that corresponds to the critical (lowest) signal y_D, or $y(z)_D$, discernible from the baseline or no signal y_0. By Kaiser's definition, this signal is

$$y_D = y_0 + 3\sigma_{y0} \qquad (6.28)$$

In a straight-line calibration $y = a + bx$, where $a = y_0$ is the blank signal and $b = S$ the sensitivity we have

$$y_D = y_0 + Sx_D \qquad (6.29)$$

that we can equate to Eq. (6.28) and obtain the limit of detection

$$x_D = \frac{3\sigma_{y0}}{S} = \frac{3\sigma_{y0}}{b} \qquad (6.30)$$

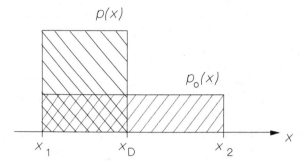

Figure 6.15. Uniform distributions for a priori and a posteriori uncertainty

Two possibilities are considered in trace analysis:

1. *No signal appears.* The analyte content is below the detection limit, $\mu < x_D$; the information gain of the result then is

$$I(p, p_0) = \mathrm{lb}\, \frac{x_2 - x_1}{x_D - x_1} \approx \mathrm{lb}\, \frac{x_2}{x_D} \qquad (6.31)$$

where $x_2 - x_1$ is the expectation range, x_2 the maximum analyte content expected in advance, x_1 the minimum content ($x_1 \ll x_2$, $x_1 \ll x_D$, frequently $x_1 \to 0$), x_D the detection limit for the method and analyte concerned; see Fig. 6.15.

In practice we always have $x_2 > x_D$ because there is no sense in choosing to determine the analyte content $\mu_0 \leq x_D$ by a method that does not work; for $x_2 \leq x_D$ the information gain would be $I(p, p_0) \leq 0$.

2. *A signal appears.* The analyte content is $\mu \geq x_D$, but the signal-to-noise ratio is low, and the probability distribution of the signal intensity is asymmetrical. The distribution of the trace content then cannot be regarded as symmetrical (as demonstrated by Eckschlager and Stepánek 1978, 1981). Rather it can be described by a shifted logarithmic-normal distribution [compare Eq. (3.27)] or a truncated normal distribution [see Eq. (3.28) and Fig. 3.4]. The probability density of the logarithmic-normal distribution shifted to

the detection limit $x_D \geq 0$ is

$$p(x) = \begin{cases} 0 & x \leq x_D \\ \dfrac{1}{(x-x_D)\sigma\sqrt{2\pi}} \exp\left[-\dfrac{1}{2}\left(\dfrac{\ln(x-x_D)-\mu}{\sigma}\right)^2\right] & \\ & x > x_D \geq 0 \end{cases}$$

(6.32)

After setting $\mu = \ln(q \cdot x_D)$, the information gain of trace analysis takes the form

$$I(p,p_0) = \mathrm{lb}\, \frac{x_2}{x_D}\frac{1}{\sigma q\sqrt{(2\pi e)}}$$

(6.33)

The probability density for a normal distribution truncated at the point $z_D = (x_D - \mu)/\sigma$ is

$$p(x) = \begin{cases} 0 & x \leq x_D \\ \dfrac{1}{[1-\phi(z_D)]\sigma\sqrt{(2\pi)}}\exp\left[-\dfrac{1}{2}\left(\dfrac{x-\mu}{\sigma}\right)^2\right] & x > x_D \geq 0 \end{cases}$$

(6.34)

and the information gain of this distribution is

$$I(p,p_0) = \mathrm{lb}\, \frac{x_2}{\sigma\sqrt{(2\pi e)}} - \mathrm{lb}[1-\phi(z_D)] + \kappa\frac{z_D\varphi(z_D)}{1-\phi(z_D)}$$

(6.35)

where $\varphi(z_D)$ is the value of the frequency function and $\phi(z_D)$ of the normal distribution function for z_D. Equation (6.35) gives a particular case of Eq. (3.28) for $z_1 = z_D$ and for z_2 so large that $\phi(z_2) \approx 1$ and $\varphi(z_2) \approx 0$. It is clear that as the analyte content increases, the $\varphi(z_D)$ and $\phi(z_D)$ in Eq. (6.35) become very small,

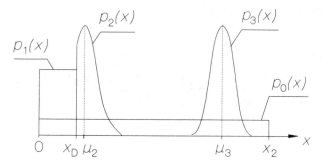

Figure 6.16. Several normal a posteriori distributions where there is uniform a priori uncertainty

and for $z_D \leq -3$ the relation transforms into

$$I(p, p_0) = \text{lb} \, \frac{x_2}{\sigma\sqrt{(2\pi e)}} \tag{6.36}$$

that is, into the relation for the normal a posteriori distribution and the uniform a priori distribution with $x_1 = 0$. Hence the information gain depends on both the analyte content μ and the detection limit x_D. This situation is represented in Fig. 6.16 and by the following example:

Now let us consider a method of trace analysis whose detection limit is $x_D = 4$ ppm and standard deviation $\sigma = 0.2$ ppm (constant across the entire region of applicability of the method). We expect these to be no more than $x_2 = 10$ ppm analyte.

First, we assume that there will be no signal, as in distribution $p_1(x)$ of Fig. 6.16. The information content of the result for $\mu < x_D$ is

$$I(p, p_0) = \text{lb} \, \frac{10}{4} = 1.32 \, \text{bit}$$

Second, we obtain a result $\mu = 4.125$ ppm. This situation is characterized in Fig. 6.16 by the distribution $p_2(x)$. So we have

$z_D = [(4.000 - 4.125)/0.2] = -0.625$ and

$$I(p, p_0) = \text{lb}\frac{10}{0.2 \times 4.133} - 0.7213\left(\frac{0.625 \times 0.328}{1 - 0.266}\right)$$
$$- \text{lb}(1 - 0.266) = 3.84 \, \text{bit}$$

Finally, we consider the case $\mu \gg x_D$, as shown by the a posteriori distribution $p_3(x)$ in Fig. 6.16. The information gain amounts to

$$I(p, p_0) = \text{lb}\frac{10}{0.2 \times 4.133} = 3.60 \, \text{bit}$$

A quantity that is important in trace analysis is the *limit of determination* (*quantitation*)

$$x_Q = w\sigma_{x0} = \frac{w\sigma_{y0}}{b} \tag{6.37}$$

It is analogous to the detection limit. Yet value w proposed by some authors is not the same. Frequently we find $w = 10$; Currie (1968) interprets this by the fact that at this value the relative precision of the analyte result determined at concentration x_Q is characterized by a value $\sigma_{rel} = 10\%$, which we consider acceptable. Still, unlike the detection limit, it is not possible to characterize the limit of determination by a concrete, generally valid value such as $w = 10$.

Currie's definition led to the notion of the limit of determination as the lowest result that can be determined with a sufficient precision. What is "sufficient" depends on individual calls for precision which in most cases is decided by the trace-analytical problem. Should a relative precision of about 33% be sufficient, then with $w = 3$ limit of determination becomes equal to the detection limit.

In their 1979 monograph Liteanu and Rica use other precision characteristics as well, such as entropy. But even they fail to take into account the accuracy of the result. In Bayermann's monograph (1982) on trace analysis of organic substances, the limit of determination is defined as the lowest result that is sufficiently precise and accurate to be deemed a satisfactory estimate of the true analyte content.

A positive information gain can be taken as the condition of a sufficiently precise and accurate result. The information gain is zero if

$$\frac{x_2 - x_1}{\sigma} = \sqrt{(2\pi e)} \, \exp\left[\frac{1}{2}z(\alpha)^2\right] \qquad (6.38)$$

where $z(\alpha)$ is the critical value of the normal distribution for the $(1 - \alpha)$ level at which the mean error is statistically significant. The determination of the analyte concentration $\mu < x_2$, $\mu \in \langle x_0, x_3 \rangle$, $x_3 \le x_2$, then always shows a positive information gain. So, after setting $x_1 = x_0$, the *lower limit of the nonzero information content x_1* is

$$x_I = x_0 + \sigma\sqrt{(2\pi e)} \, \exp\left[\tfrac{1}{2}z(\alpha)^2\right] = x_0 + A(\alpha)\sigma \qquad (6.39)$$

This is an analogue to the definition of the limit of determination given by Eq. (6.37) for $w = A(\alpha)$ and $x_0 = 0$. In practice, however, it is difficult to determine whether an error is really a bias, particularly in the determination of very low contents. For this reason Eckschlager (1988) suggests that the value of w should be determined as follows:

$$w = \begin{cases} 10 & \text{for } 0 \le \dfrac{\delta}{\sigma} < 1.33 \\[2ex] A(\alpha) & \text{for } 1.33 \le \dfrac{\delta}{\sigma} \le 1.96 \end{cases} \qquad (6.40)$$

The case where $(\delta/\sigma) > 1.96$ would never occur in practice. It says that the results carry an error that is statistically significant at a level $(1 - \alpha) \ge 0.95$. Such an error would always be eliminated or at least reduced as one modifies the working procedure, the calibration procedure, and so on. The $A(\alpha)$ values have been tabulated by Eckschlager (1988) for $(1 - \alpha) = 0.95$ and $1.33 \le \delta/\sigma \le 1.96$, e.g., $A(0.05) = 28.21$.

Given detection limit x_D and the limit of nonzero information content x_1, we can divide the entire possible trace analyte content

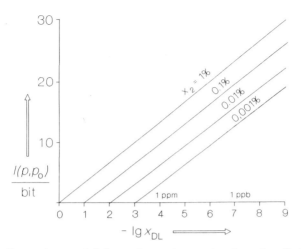

Figure 6.17. Dependence of information gain on the detection limit when the presence of an analyte cannot be detected

region into three ranges:

1. $\mu < x_D$. The presence of analyte cannot be detected by the given analytical method. We only know that potentially $0 \leq \mu < x_D$. The information content is given by Eq. (6.31); the increase of information content in that case with decreasing detection limit can be seen in Fig. 6.17.

2. $x_D \leq \mu < x_I$. The presence of analyte can be detected but probably not quantitatively determined, since the result can have a zero information gain. The information content is given by Eq. (6.33) or (6.35).

3. $\mu \geq x_I$. Analyte content can be quantitatively determined, and the results, unless involving an error that is statistically significant at level $(1 - \alpha) > 0.95$, always have a positive information contribution which is given by Eq. (6.36).

Thus for different methods of determining analytes in various materials, or for different procedures, calibration methods, or for different sensitivities of instrumentation, the x_D and x_I values will be different, as will be the regions within which only qualitative or quantitative analysis is possible.

As Currie observed in 1968, we must distinguish between the "a priori" value (the mean or limiting value of the metrological characteristics valid for the analytical method) and its "a posteriori" value (the true, actual value for the procedure applied to the detection or determination of the analyte in the sample of a given matrix including sampling, sample handling, calibration, calculation of the result, and so on). The a priori values of the characteristics are involved, for instance, in the choice of the optimum value, and they are based on parameters of the theoretical distribution, which is usually normal. The a posteriori, actual values, are based on parameter estimates and hence on random quantities, and they are important in the optimization of procedures and in quality data assurance. Everywhere it is important that the results be processed into relevant information.

All that has been said with respect to identification, qualitative and quantitative analysis of higher concentrations, also applies to trace analysis. However, some additional, largely unfavorable circumstances must be considered in the latter case:

1. For contents lower than the detection limit or the nonzero information limit, only a qualitative (or semiquantitative) analysis can be carried out at best, despite our wanting to know "what" and "how much." This is controlled by the type of method, instrument, procedure, and so on.

2. Information gain must be expressed for different a posteriori probability distributions by the relationship between the analyte content and the detection limit, the limit of determination, or the limit of nonzero information according to the method used.

3. Metrological backup of calibration is more difficult for trace analysis than for the determination of major components, but it is also considerably more important. It is often necessary to subtract the blank value, which itself can be rather variable. Contamination and loss of analyte are difficult to predict even in a well-equipped "clean" laboratory.

4. The relative standard deviation of the results is usually large, and only a large bias δ can be identified as statistically significant. Therefore it is advisable to replicate results more than is usual in the analysis of major components.

6.5.3   Practical Applications

A good example of the practical application of trace analysis is determining lead traces in soil. In our example the samples were taken in the surroundings of a superhighway at different distances from the road. After drying, the sandy soil samples were analyzed by optical emissions spectroscopy (OES) by direct evaporation and arc excitation. Because of the varying distances the values were expected to range from $x_1 < x_D$ up to $x_2 = 100$ ppm; the relevant parameters of the analytical procedure were $x_D = 1$ ppm, $x_I = 4$ ppm; for the analytical error we will assume two cases: (1) a constant relative error $\sigma_{rel} = 0.15$ and (2) a constant absolute error $\sigma = 0.1$ ppm, where the case (1) is the more realistic one.

Fig 6.18 shows, as in Eq. (6.31), that the information content is constant up to the detection limit for all cases in which no lead is found. In cases where Pb is found, it can be seen that for a constant error $\sigma = 0.1$ ppm and for the content of Pb higher than about 1.5 ppm the information content is constant; namely $I(p, p_0) =$ lb$[100/(0.1 \times 4.133)] = 7.92$ bit. In contrast, in case a constant relative error the absolute error term increases as in Eq. (6.36), and therefore the information content decreases from 6.69 bit for 1.5 ppm Pb found to 4.00 bit for 10 ppm Pb. Depending on the size of the error, the information content may be higher if the analyte is

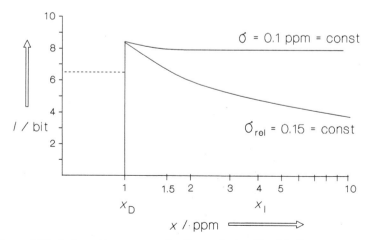

Figure 6.18. Information content for the analysis of lead in the ppm range

not found than if the analyte is detected. This situation can be seen from the lower curve in Fig. 6.18. It agrees with our analytical experience that the result "Pb content is lower than 1 ppm" may be more interesting and important than the result (10 ± 1.5) ppm Pb is in some cases.

6.6 EFFECT OF PRECISION AND ACCURACY OF RESULTS ON INFORMATION GAIN

6.6.1 General Aspects

Important in analytical practice is the dependence of $I(r; p, p_0)$ on the precision of the results, characterized by the random error σ, and on their accuracy, expressed by the bias $\delta = |X - \mu|$. The dependence of $I(r; p, p_0)$ on σ is given in Fig. 6.19. As the figure demonstrates, although more precise results for $\delta = 0$ exhibit higher information gain than less precise results, the $I(r; p, p_0)$ value decreases with increasing δ much faster for more precise results than for less precise results.

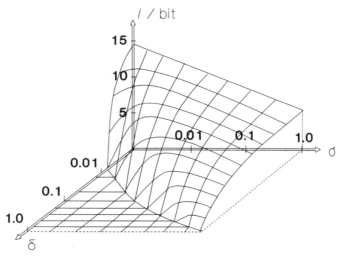

Figure 6.19. Dependence of information gain on random error σ and systematic error δ

The value of δ inserted in relations expressing the information gain is any value of the error $\delta = X - \mu$. In practice, however, it is very important to discriminate between a statistically insignificant error $\delta \leq z(\alpha)\sigma$ and an error statistically significant at a level of significance $(1 - \alpha)$, that is, for $\delta > z(\alpha)\sigma$, where $z(\alpha)$ is the critical value for the level of significance $(1 - \alpha)$ at which the error is statistically significant.

The information gain of results involving such an error then is given according to Eq. (6.36) as

$$I(r; p, p_0) = \text{lb} \frac{x_2 - x_1}{\sigma\sqrt{(2\pi e^k)}} - \kappa z(\alpha)^2 \qquad (6.41)$$

and according to Eq. (6.39) as

$$I(r; p, p_0) = \text{lb} \frac{\sigma_0}{\sigma} + \kappa\left[\left(\frac{\mu - \mu_0}{\sigma_0}\right)^2 + k\frac{\sigma^2 - \sigma_0^2}{\sigma_0^2}\right] - \kappa z(\alpha)^2$$

$$(6.42)$$

In general, for both cases we can write

$$I(r; p, p_0) = \kappa A - \text{lb}\, R - \kappa z(\alpha)^2 = I(p, p_0) - \kappa z(\alpha)^2 \quad (6.43)$$

where $I(p, p_0)$ is the information gain according to Eq. (6.16) and A characterizes the divergence of the a priori and a posteriori distributions and thus depends on k, or can involve the measure of surprise D_0^2. Hence it characterizes the well-known fact that the significance of the result of an experiment is not given solely by its precision and accuracy. $I(r; p, p_0)$ in Eq. (6.43) is the information gain attainable at $\delta = 0$.

The dependence of $I(r; p, p_0) = I_0$ on $z(\alpha)$ is shown at the right-hand side of Fig. 6.20. It is apparently more convenient to regard the dependence of the information gain on the statistical significance than on the magnitude of error δ. L_1 and L_2 are the lower and upper limits, respectively, of the uncertainty interval of the μ value.

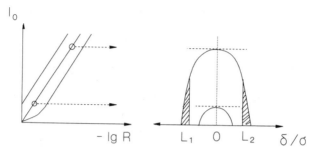

Figure 6.20. Dependence of information gain on $-\log R$ and δ/σ

In practice, the true value of error is largely unknown and can be determined only exceptionally. Mostly, however, metrological provisions can be taken for an actual error that is statistically insignificant at a level $(1 - \alpha)$. Then we only know that for the a priori rectangular distribution,

$$\text{lb } \frac{x_2 - x_1}{\sigma\sqrt{(2\pi e^k)}} - \kappa z(\alpha)^2 \leq I(r; p, p_0) \leq \text{lb } \frac{x_2 - x_1}{\sigma\sqrt{(2\pi e^k)}} \quad (6.44)$$

and for the a priori normal distribution,

$$\text{lb } \frac{\sigma_0}{\sigma} + \kappa\left[\left(\frac{\mu - \mu_0}{\sigma_0}\right)^2 + k\frac{\sigma^2 - \sigma_0^2}{\sigma_0^2}\right] - \kappa z(\alpha)^2$$

$$\leq I(r; p, p_0) \leq \text{lb } \frac{\sigma_0}{\sigma} + \kappa\left[\left(\frac{\mu - \mu_0}{\sigma_0}\right)^2 + k\frac{\sigma^2 - \sigma_0^2}{\sigma_0^2}\right] \quad (6.45)$$

In general, the interval within which there lies the true information gain which can at most involve an error statistically significant at the level $(1 - \alpha)$ is

$$I(r; p, p_0)_0 - \kappa z(\alpha)^2 \leq I(r; p, p_0) \leq I(r; p, p_0)_0 \quad (6.46)$$

where $I(r; p, p_0)_0$ is the information gain for $\delta = 0$ proved experimentally.

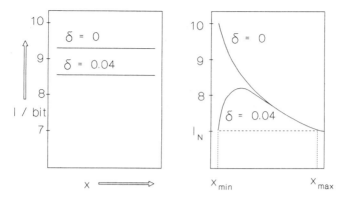

Figure 6.21. Dependence of information gain on μ for the homoscedastic (*left*) and heteroscedastic (*right*) cases

An important consideration for the width of the intervals (6.44) to (6.46) is whether the standard deviation is constant over the range $\langle x_1, x_2 \rangle$ (i.e., the calibration dependence is homoscedastic) or whether the standard deviation varies with μ (i.e., the calibration dependence is heteroscedastic). Changes in $I(r; p, p_0)$ in dependence on μ for the two cases are shown in Fig. 6.21.

It is clear that in the former case the width of the interval (6.46) is constant and only determined at the level where the error δ is statistically significant. For heteroscedastic dependence the width varies with μ and can drop appreciably as μ approaches x_{min}. The effect of homoscedastic and heteroscedastic calibration dependence on the information gain of the quantitative analysis has been considered by Eckschlager and Fusek (1988).

While the upper limit of interval (6.46) is always positive, the lower limit can be negative. Although classical information theory does not define negative information, we can — with respect to the pragmatic meaning of the analytical result—interpret $I(r; p, p_0) < 0$, as a case where incorrect results *misinform* us rather than bring relevant information. To obtain a positive information contribution, we must have, for the significance level $(1 - \alpha) = 0.95$ with $z(0.05) = 1.96$, $I(r; p, p_0) \geq \kappa(1.96)^2 = 2.77$ bit by Eqs. (6.41) and (6.42). This condition must be fully satisfied in the quantitative determination of major components; in ultra-trace analysis, however, the attainment of a positive information contribution may not be a matter of course.

The extended divergence measure can serve as a "warning" where results involve a bias, through the situation hardly occurs in practice. The measure is of use as a characteristics for assessing to what extent the analytical results approach the truth (i.e., the true, objectively existing analyte content in sample), since $I(r; p, p_0)$ can also attain negative values. The problem of the truth and convergence to it in analytical results are discussed by Malissa (1988, 1989).

In considering the accuracy of quantitative results, it is appropriate to devote some attention to signal decoding. Signal interpretation must meet two objectives:

1. Yield all the information relevant to the solution of the problem.
2. Yield the required information about chemical composition with the lowest possible a posteriori uncertainty; that is, it should represent the highest possible information gain.

In performing the calibration, the most suitable model (i.e., concerning the shape of the calibration dependence) must first be chosen. The linear dependence $y = a + bx$ is the one most frequently used, and it also corresponds to many actual physical relationships such as Lambert-Beer's and Ilkovic's equation. In chemical analysis we have $y = bx$, where b is the stoichiometric equivalent (hence a constant) and the relation between the standard deviation of the result σ and the standard deviation of the signal σ_y, which is given as

$$\sigma = \frac{\sigma_y}{b\sqrt{n}} \tag{6.47}$$

In instrumental analysis, even if the dependence of y on x, which is found empirically by calibration, is linear, the relationship between σ and σ_y also depends on the conditions under which the calibration is carried out. For instance, if a calibration curve set up of m

points is used and the signal measurement is n-fold repeated, we have

$$\sigma = \frac{\sigma_y}{b} \sqrt{\frac{1}{n} + \frac{1}{m} + \left(\frac{\sigma_b}{b\sigma_y}\right)^2 y^2} \qquad (6.48)$$

The two equations demonstrate how important for the information gain, achieved through results obtained by decoding a signal, is the sensitivity, which is characterized by the constant or slope of the calibration straight line b. The effect of sensitivity on the information gain is particularly marked in multicomponent and trace analysis.

6.6.2 Reporting Analytical Results and Information Gain

When reporting results of quantitative analysis, we face the problem of how to express their uncertainty. One of the many was to express this is by point characteristics. If the uncertainty is only due to imprecision, it suffices to report the absolute or relative standard deviation. This way of characterizing results, however, does not take into account inaccuracy, which from the information contribution perspective is more serious than imprecision.

The various proposed point characteristics of precision and accuracy, such as *total error* suggested by McFarren et al. (1970) and its modifications, are defined more from practical than from theoretical point of view. Yet, these point characteristics are not used in practice. The measure of accuracy or overall accuracy proposed by Eckschlager and Stepánek (1980), although theoretically well founded, have not found use in ordinary practice either. Apparently point characteristics are not well suited to expressing the uncertainty of analytical results.

Quite common is the use of interval characteristics whereby the result is reported as μ in the form $L_1 \le \mu \le L_2$, where L_1 and L_2 are the lower and upper limits, respectively, of the interval whose width characterizes the limits of uncertainty of the μ value. This uncertainty, though, must be defined uniquely, and the conditions for which the relation expressing its magnitude has been derived must be satisfied. If only the uncertainty due to the imprecision of normally distributed results is considered, the confidence interval concept is commonly used; for this we have

$$P\left[\bar{x} - \frac{st(\alpha, f)}{\sqrt{n}} \leq \mu \leq \bar{x} + \frac{st(\alpha, f)}{\sqrt{n}}\right] = (1 - \alpha) \quad (6.49)$$

The confidence level $(1 - \alpha)$ is basically arbitrarily chosen, although the convention is to use $(1 - \alpha) = 0.95$. The solution for the case where the lower limit obtained in quantitative analysis is lower than 0% or higher than 100% is given in the monograph by Doerffel and Eckschlager (1981, ch. 10). The determination of the confidence interval for results with the lognormal distribution is similar; it is performed with logarithms of the results. The interval, however, is asymmetrical about μ.

Expressing the confidence interval of results carrying a bias δ is associated with some problems. Since the sign of the bias is always known, the proposal by Grabe (1986) that the confidence interval should be expressed as

$$L_{1,2} = \bar{x} \pm \left[\frac{st(\alpha, f)}{\sqrt{n}} + \delta\right]$$

cannot be regarded appropriate; addition of δ to the expected value of μ and reporting the interval that obeys for $\delta \neq 0$,

$$P\left[\bar{x} - \delta - \frac{st(\alpha, f)}{\sqrt{n}} \leq \mu \leq \bar{x} + \delta + \frac{st(\alpha, f)}{\sqrt{n}}\right] > (1 - \alpha)$$

however, is not suitable either because for this interval we actually do not know the confidence level. The confidence interval is expressed by the noncentral t-distribution for $\delta \geq 0$ as

$$P\left[\bar{x} - \frac{st(\alpha, f, \delta)}{\sqrt{n}} - 2\delta \leq \mu \leq \bar{x} + \frac{st(\alpha, f, \delta)}{\sqrt{n}}\right] = (1 - \alpha)$$

$$(6.50a)$$

or for $\delta < 0$ as

$$P\left[\bar{x} - \frac{st(\alpha, f, \delta)}{\sqrt{n}} \leq \mu \leq \bar{x} + \frac{st(\alpha, f, \delta)}{\sqrt{n}} - 2\delta\right] = (1 - \alpha)$$

$$(6.50b)$$

can be determined as given by Eckschlager et al. (in Wilson and Wilson vol. 18, 1983). It is asymmetrical about x but symmetrical about μ.

Reporting the information contribution of results poses problems similar to those encountered in the presentation of the results themselves. In view of what has been said, we actually only know the interval

$$I(r; p, p_0)_0 - \kappa z(\alpha)^2 \leq I(r; p, p_0) \leq I(r; p, p_0)_0 \quad (6.51)$$

whose width for given σ and δ is determined by the level $(1 - \alpha)$ at which the error is statistically significant. Hence this is an analogy of the confidence interval.

A comment on the interval estimates: expressing the interval—symmetrical or asymmetrical—we tacitly assume the rectangular distribution of the quantity in question within the limits of L_1 to L_2, which is actually a simplification. The width of the confidence interval and of the interval (6.51) in dependence on the level $(1 - \alpha)$ is schematically shown in Fig. 6.22.

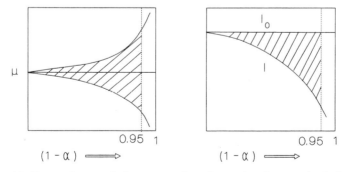

Figure 6.22. Dependence of the uncertainty intervals of μ and of I_0 on the confidence level $(1 - \alpha)$

6.6.3   Practical Applications

In practice, the true values of the a posteriori distribution parameters are usually unknown, and their estimates can only be derived from the data set $(x_1, x_2, \ldots, x_n)$, where the number n is usually so low that the estimates can differ quite markedly from the true values. Therefore the information gain *estimate* concept has been introduced.

For the uniform a priori distribution, the estimate of information gain is

$$\hat{I}(r; p, p_o) = \text{lb}\frac{(x_2 - x_1)\sqrt{n}}{2st(\alpha^*, f)} - \frac{1}{2}n\,\text{lb}\,e\left(\frac{\delta}{s}\right)^2$$

$$= \text{lb}\frac{(x_2 - x_1)\sqrt{n}}{2st(\alpha^*, f)} - \kappa t(\alpha, f)^2 \qquad (6.52)$$

This relation derives from Eq. (6.41) by substitution of $s/\sqrt{n}$ for σ; the expression $\sqrt{(2\pi e^k)}$ must be replaced with $2t(\alpha^*, f)$ and the expression $z(\alpha)^2$ with $t(\alpha, f)^2$. The term $t(\alpha^*, f)$ is the critical value of the t-distribution for $f = n_s - 1$; n_s is the number of those replicate determinations from which the standard deviation estimate has been calculated, and n is the number of actual replicate determinations. It is clear that $n_s \geq n$.

The α^* value is chosen so that $t(\alpha^*, f_{max}) \approx \frac{1}{2}\sqrt{(2\pi e^k)}$; the α value is given by the level at which the bias δ is statistically significant. Since we can take, for instance, $f_{max} = \infty$, we often use $\alpha^* = 0.039$ for which $t(0.039, \infty) = \frac{1}{2}\sqrt{(2\pi e)} = 2.066$. Alternatively, in practice, we consider a data set with no more than $n = 25$ as tractable; we choose $\alpha^* = 0.05$, for which $t(0.05, 24) = 2.066$, and so on. The $t(0.05, f)$ values are tabulated in almost each mathematical statistics textbook, and the $t(0.039, f)$ values are given in the monograph by Eckschlager and Stepánek (1979, p. 112).

In practice, it is sufficient to use simple approximations of values, such as

$$t(0.039, f) \approx 2 + \frac{4}{f}$$

or

$$t(0.05, f) \approx 2 + \frac{2.5}{f}$$

which are applicable at $4 \leq f \leq 20$. Equation (6.52) can be looked upon not only as an estimate of the information gain expressed in divergence terms but also, in the sense of Brillouin's information content concept, as a reduction of the interval $(x_2 - x_1)$ to an interval whose width is $2st(\alpha^*, f)/\sqrt{n}$; the details on Brillouin's measure, which, however, has not found wide application in analytical practice, can be found in the paper by Eckschlager (1977).

For an illustration of the application of Eqs. (6.41) and (6.52), let us determine the information gain of the determination of 0.05 to 6.00% Mn in steel. If $\sigma = 0.006\%$, we have

$$I(p, p_0) = \text{lb} \frac{6.00 - 0.05}{0.006 \times 4.1327} = 7.91 \, \text{bit}$$

This is the case of using "constant" to yield results that are accurate.

In reality, however, the σ value is unknown. Therefore we determine the estimate $s = 0.0081$ from $n_s = 8$, taking 24 for f_{max} and performing $n = 2$ repetitions. Then we have $t(0.05, 7) = 2 + 2.5/7 = 2.36$; the tabulated value is $t(0.05, 7) = 2.365$ and

$$\hat{I}(p, p_0) = 7.78 \, \text{bit}$$

If the number of replications is increased, we have

$$\hat{I}(p, p_0) = 8.07 \, \text{bit} \qquad \text{for } n = 3$$

$$\hat{I}(p, p_0) = 8.28 \, \text{bit} \qquad \text{for } n = 4$$

A further increase in the number of replications has an even smaller effect.

All this only holds true if the results are accurate. Now, assume that by verifying the method with a reference material ($k = 0.0625$), we find that the results involve a bias of $\delta = 0.008\%$. Since

$\sqrt{(2\pi e^{0.0625})} = 2.586$, which corresponds to $t(0.2, \infty)$, we have $t(0.2, 7) = 1.415$ and

$$\hat{I}(r; p, p_0) = \text{lb} \frac{(6.00 - 0.05)\sqrt{2}}{2 \times 0.0081 \times 1.415} - 2 \times 0.7214 \left(\frac{0.0080}{0.0081} \right)^2$$

$$= 7.11 \text{ bit}$$

The bias can be eliminated by subtracting the blank value. Then the s increases from 0.0081 to $0.0081\sqrt{2} = 0.01145$ and the information gain is

$$\hat{I}(r; p, p_0) = \text{lb} \frac{(6.00 - 0.05)\sqrt{2}}{2 \times 0.01145 \times 1.415} = 8.02 \text{ bit}$$

It is clear that the information gain is affected by the number of repeated determinations, the verification of the method's accuracy, and the elimination of the bias (if applied). This value can be lower or higher than the "constant" theoretical value calculated from the expected value σ and for $\delta = 0$.

A certain risk may result from the use of unsuitable reference materials (e.g., as when a larger error σ_r is caused by the inhomogeneity of the material). For large values of $k = (\sigma_r/\sigma)$ the information gain $I(r; p, p_0)$ can take a negative value. In such situations —such as given by the relations $(x_2 - x_1) = 6\sigma$ and $k \geq 1.74$ or for $(x_2 - x_1) = 10\sigma$ and $k \geq 2.76$, where in either case $I(r; p, p_0) \leq 0$—the analytical results *misinform* us. "Classical" information theory does not allow negative values of information gain, so Eckschlager and Stepánek (1985) introduced the "null information content" whereby

$$I = \begin{cases} I(r; p, p_0) & \text{for } I(r; p, p_0) > 0 \\ 0 & \text{for } I(r; p, p_0) \leq 0 \end{cases} \tag{6.53}$$

The case $I = 0$ is called the "null information content" regardless of $I(r; p, p_0)$ being able to assume even a large negative value.

In practice, it is important to know that the meaning of the value $I = 0$ may be somewhat different, namely $I(r; p, p_0) = 0$ which means that "the information gain is zero," or "no information contribution," and $I(r; p, p_0) < 0$ which means that misinformation is present to some degree. In either case it is critical to check the source of the inaccuracy, as well as what can be done to improve it.

REFERENCES

Bayermann, K. 1982. *Organische Spurenanalyse*. Thieme, Stuttgart.

Becker, D. 1987. J. Radioanal. Nucl. Chem. **113**, 5.

Brillouin, L. 1963. *Science and Information Theory*. Academic Press, New York.

Currie, L. A. 1968. *Anal. Chem.* **40**, 586.

Currie, L. A. 1993. Position paper on nomenclature for calibration in analytical chemistry. Internal paper of the IUPAC Commission V.1 unpublished.

Danzer, K. 1990. *Fresenius J. Anal. Chem.* **337**, 794.

Danzer, K. 1994. *Fresenius J. Anal. Chem.*, submitted.

Danzer, K., Eckschlager, K., and Matherny, M. 1989. *Fresenius Z. Anal. Chem.* **334**, 1.

Danzer, K., Eckschlager, K., and Wienke, D. 1987. *Fresenius Z. Anal. Chem.* **327**, 312.

Danzer, K., Schubert, M., and Liebich, V. 1991. *Fresenius J. Anal. Chem.* **341**, 511.

Danzer, K., Than, E., Molch, D., and Kochler, L. 1987. *Analytik— Systematischer Überblick*. Wissenschaftliche Verlagsgesellschaft, Stuttgart.

Danzer, K. 1973. *Z. Chem.* **13**, 229.

Doerffel, K. and Eckschlager, K. 1981. *Optimale Strategien in der Analytik*. Deutscher Verlag für Grundstoffindustrie, Leipzig.

Doerffel, K., Eckschlager, K., and Henrion, G. 1990. *Chemometrische Strategien in der Analytik*. Deutscher Verlag für Grundstoffindustrie, Leipzig.

Doerffel, K., and Hildebrandt, W. 1969. *Wiss. Z. Techn. Hochsch. Leuna-Merseburg* **11**, 30.

Eckschlager, K. 1971. *Collect. Czech. Chem. Commun.* **36**, 3016.

Eckschlager, K. 1975. *Fresenius Z. Anal. Chem.* **277**, 1.

Eckschlager, K. 1977. *Collect. Czech. Chem. Commun.* **42**, 225.

Eckschlager, K. 1988. *Collect. Czech. Chem. Commun.* **53**, 1647.

Eckschlager, K., and Fusek, J. 1988. *Collect. Czech. Chem. Commun.* **53**, 3021.

Eckschlager, K., and Stepánek, V. 1978. *Mikrochim. Acta* **I**, 107.

Eckschlager, K., and Stepánek, V. 1979. *Information Theory as Applied to Chemical Analysis.* Wiley, New York.

Eckschlager, K., and Stepánek, V. 1980. *Collect. Czech. Chem. Commun.* **45**, 2516.

Eckschlager, K., and Stepánek, V. 1981. *Mikrochim. Acta* **II**, 143.

Eckschlager, K., Stepánek, V., and Danzer, K. 1990. *J. Chemometrics* **4**, 195.

Eckschlager, K. et al 1983. *Application of Computers in Analytical Chemistry*, in Wilson and Wilson eds. Comprehensive Analytical Chemistry, Vol 18; Elsevier, Amsterdam.

Fujimori, T., Miyazu, T., and Ishikawa, K. 1974. *Microchim. J.* **19**, 74.

Grabe, M. 1986. *Metrologia* **23**, 213.

Heydorn, K. 1988. *J. Res. NBS* **93**, 479.

International Union of Pure and Applied Chemistry, Analytical Chemistry Division. 1977. *Compendium of Analytical Nomenclature.* Prepared for publication by Irving, H. M. N. H., Freiser, H., West, T. S. Pergamon Press, Oxford, 15.

Kaiser, H. 1965. *Fresenius Z. Anal. Chem.* **209**, 1.

Kateman, G. 1986. *Anal. Chim. Acta* **191**, 215.

Liteanu; C., and Rica, I. 1979. *Statistical Theory and Methodology of Trace Analysis.* Ellis Horwood, Chichester.

Malissa, H. 1988. *Fresenius Z. Anal. Chem.* **331**, 236.

Malissa, H. 1989. *Fresenius Z. Anal. Chem.* **333**, 285.

McFarren, E. F., Lishka, R. J., and Parker, J. H. 1970. *Anal. Chem.* **42**, 358.

Musil, J. 1986. *Chem. Listy* **80**, 1233.

Obrusnik, I., and Eckschlager, K. 1993. *J. Radioanal. Nucl. Chem.* **169**, 347.

Rasberry, S. D. 1988. *J. Res. NBS* **93**, 213.

Tölg, G. 1976. *Naturwissenschaften* **63**, 99.

Tölg, G. 1987. *Analyst* **112**, 365.

Tölg, G. and Tschöpel, P. 1987. *Anal. Sci.* **3**, 199.

Versieck, J., Vanballenberghe, L., de Kesel, A., Hoste, J., Wallaeys, B., Vandenhaute, J., Baeck, N., Steyaert, H., Byrne, A. R., Sunderman, Jr., F. W. 1988. *Anal. Chim. Acta* **204**, 63.

Woschni, E. -G. 1970. *Information und Automatisierung*. Verlag Technik, Berlin.

Woschni, E. -G. 1972. *Meßdynamik*. Hirzel, Leipzig.

Ziessow, D. 1973. *On-line Rechner in der Chemie*. de Gruyter, Berlin.

CHAPTER

7

MULTICOMPONENT ANALYSIS

Most of the present methods of qualitative, identification, and quantitative analysis are instrumental enabling multicomponent analysis. With multicomponent analysis we can obtain simultaneously answers to the questions "what" and "how much" for several components from two-dimensional analytical information, as shown in Fig. 7.1.

The information obtained by a two-dimensional analysis according to Eq. (3.2), where $M = \Sigma I_i$, however, represents the sum of information contributions of the individual determinations. This information is valid only if the results are independent. The independence of results cannot be assumed always in multicomponent analysis. The reasons are the following:

1. Some of the components may be interconnected in that some may share pollution sources, as is true of impurities in materials or in the environment and some may interact with each other in the analytical system.
2. If all the results come from one sample, then the sampling, sample decomposition, sample preparation, and various separations are all the same, and so there will be a "common" bias.

As a result the actual amount of information is usually lower in analytical practice:

$$M \leq \sum_{i=1}^{n} I_i \qquad (7.1)$$

particularly if imperfect selectivity increases the a posteriori uncertainty of the individual determinations.

131

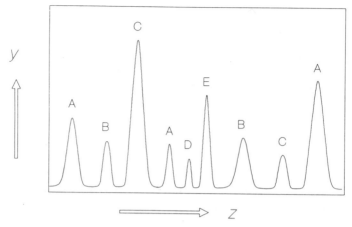

Figure 7.1. Two-dimensional analytical signal function $y = f(z)$; the different signals belong to components $A, B, \ldots, E$

In the ideal case where the prerequisites applying to all components involve the same expectation ranges, the same precision, the same a priori probabilities, and so on, the components and their signals are independent. The information amount is given by $M = \Sigma I_i$ and its maximum value by

$$M = n \cdot I_{\max} = n \cdot \text{lb } m \qquad (7.2)$$

where n is the number of components and m the number of distinguishable concentration steps (or concentration resolution power; see Doerffel and Hildebrandt 1969; Danzer 1973a, 1973b; Danzer, Eckschlager, and Matherny 1989).

In the case of correlated signals and components, the signal-to-noise ratio [see Eqs. (6.11) and (6.12)] can be derived from

$$I = \frac{1}{2}\text{lb}\frac{\sigma_S^2}{\sigma_N^2} \qquad (7.3)$$

where σ_S^2 is the variance of the signal and σ_N^2 the variance of the noise. Their quotient is the signal-to-noise power ratio. In the general case the information amount of multicomponent analysis (see Dupuis and Dijkstra 1975) is given by

$$M = \frac{1}{2} \mathrm{lb} \frac{\det|C_S|}{\det|C_N|} \tag{7.4}$$

where $\det|C_S|$ is the determinant of the variance-covariance matrix of the signal intensities and $\det|C_N|$ the variance-covariance matrix of the noise. For example,

$$C_S = \begin{pmatrix} \sigma_{11} & \sigma_{12} & \cdots & \sigma_{1n} \\ \sigma_{21} & \sigma_{22} & \cdots & \sigma_{2n} \\ \vdots & \vdots & & \vdots \\ \sigma_{n1} & \sigma_{n2} & \cdots & \sigma_{nn} \end{pmatrix} \tag{7.5}$$

where

$\sigma_{ii} = \sigma_i^2$ the variance of the signal i

$\sigma_{ij} = \sigma_i \cdot \sigma_j \cdot r_{ij}$ the covariance of the signals i and j

$r_{ij} =$ correlation coefficient of the signals i and j

If no correlation exist between the signals of different components, C_S with $\sigma_{ij} = 0$ becomes a diagonal matrix. Therefore Eqs. (7.5) and (7.4) change into Eq. (7.2) or (7.1). Correlated signals particularly occur in mass spectroscopy and atomic spectroscopy (RFA, OES) because several signals belong to the same component.

Information theory is more useful in multicomponent analysis than in single-component analysis because it enables us to include among optimization criteria selectivity of methods, relevance of results, redundancy of procedures, and the effect of signal processing, which all are highly significant in this case, in addition to the precision and accuracy.

In the instrumental methods of qualitative, identification, or quantitative multicomponent analysis, both the first stage (experimentation) and the second (evaluation) are complicated. The most important operation in the first stage is of course measurement, which involves the most important property of an instrument—its resolving power. The evaluation stage usually encompasses several

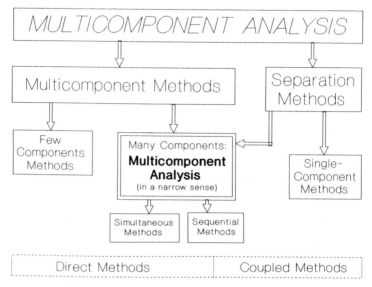

Figure 7.2. Possible multicomponent analyses

operations: signal encoding, smoothing and deconvolution, accumulation and filtering (particularly recursive as in Kalman's filtering), and so on.

7.1   METHODS OF MULTICOMPONENT ANALYSIS

There are two different strategies for applying instrumental methods in multicomponent analysis, and therefore two possible approaches to multicomponent analysis (see Fig. 7.2):

1. A direct, two-dimensional method with a sufficiently high resolution to register side by side the signals of different, largely undisturbed components.

2. A coupling technique that separates and determines the signals.

Among the methods inherently multicomponent in character—or "few component methods," also called "oligocomponent methods" —are some of less importance because they enable only the detection and determination of a few components, namely two to about five.

- *Photometry*. This classical single- or two-component method can be applied in combination with special chemometric evaluation techniques using latent variables as in PLS (partial least squares procedure according to Wold and coworkers; see Sjöström, Wold et al. 1983, Lindberg, Persson, and Wold 1983) to analyze simultaneously up to six components (organic dyes, complexes of metal ions, etc.). By using the PLS algorithm and related chemometric procedures, the lack of selectivity in the determination method is compensated by advanced procedures of evaluation (i.e., multivariate data analysis). It can be said that selectivity is moved from the realm of determination to that of evaluation. With use of chosen, nonselective reagents that react with a large number of metal ions, photometric multicomponent analysis can be realized. Take, for example, the application of ammonium tetramethylen dithiocarbamate (ATDTC) which forms complexes with about 35 elements. Seven of the elements (Mn, Cu, Ni, Fe, Co, Cr, and Pb) have been analyzed simultaneously by Schneider, Wienke, and Danzer (1987) and Jana (1989).
- *Polarography and inverse voltammetry*. This technique can analyze qualitatively and quantitatively several ions in aqueous solution simultaneously because of their different half-step potentials and heights, provided that the potentials differ sufficiently from each other.
- *Simultaneous titration*. Since a few components play a more tangential role, this is used only in special applications today.

Figure 7.3 shows some important methods of real multicomponent analysis. A distinction is made between *direct and coupled* methods, on the one hand, and multicomponent methods of recognizing organic and inorganic constituents, on the other hand.

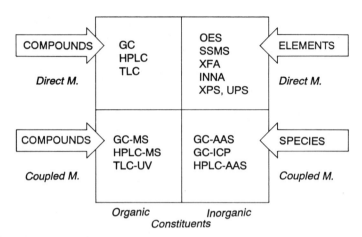

Figure 7.3. Direct and coupled methods of multicomponent analysis of organic and inorganic constituents

The most powerful methods of multielemental analysis are the following:

- *Optical emission spectroscopy (OES)*. Excited by arc, spark, glow discharge, and plasma torchs stabilized in different ways (ICP, MIP), OES distinguishes itself by high-resolution power and therefore the ability to register some hundreds of thousands of spectral lines. The signals may be recorded simultaneously (by multichannel mode) or sequentially (by scanning mode). The number of elements that can be analyzed is large: About 90 of the elements are usually accessible and the rest (nonmetals) by means of special equipment (vacuum instrument).

- *Solid state mass spectroscopy (SSMS)*. Evaporation and excitation may be carried out by means of spark sources (SSMS), glow discharge (GDMS), laser ablation (LAMS), inductively coupled plasma (ICP-MS), or ion sources (secondary ion MS, SIMS). All the elements are recognized in the spectra, as well as their isotopes, which can cause some confusion so computational corrections may be required. This problem does not exist in high-resolution mass spectrometry with double-focus-ing instruments (frequently magnetic plus electrostatic sector fields); their resolving power is in the hundreds of thousands.

- *X-ray fluorescence spectrometry* (*XRF*, *XFA*). The classical form is somewhat limited as to the number of elements—light elements up to atomic number 11 cannot usually be analyzed —and the detection limits. Recently developed *total reflection X-ray fluorescence* (TRXFA) surpasses the common XFA in detection power by up to four orders of magnitude.

- *Photoelectron spectroscopy* (XPS and UPS) and *Auger electron spectroscopy* (AES) are special multicomponent methods that not only enable us to detect a large number of elements ($Z \geq 3$) but also to recognize several oxidation and bonding states by means of characteristic chemical shifts. Although the Auger effect and X-ray fluorescence are competing methods, AES is more sensitive for light elements.

- *Neutron activation analysis*. This method is suitable for multi-element analysis only in its instrumental version, that is, by gamma spectrometry (INAA). It generally offers high-detection power, but this differs from element to element and matrix to matrix because of different activation cross sections, half-life times, and competition reactions.

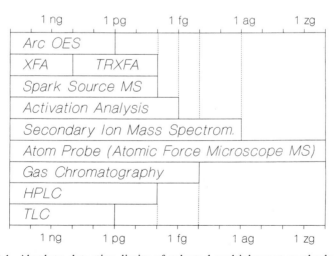

Figure 7.4. Absolute detection limits of selected multielement methods (zg: zeptogramme)

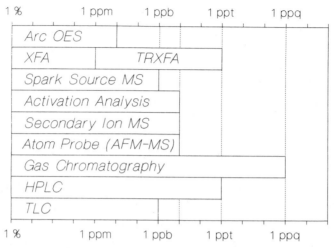

Figure 7.5. Relative detection limits of selected multicomponent methods

- *Chromatographic methods.* Generally chromatography is used in multicomponent analysis of organic compounds. Chromatographic techniques are able to separate, to identify, and to determine a large number of components, depending on their mobil and stationary phases, as well as very similar chemicals like dioxines, PHAs, PCBs, and amino acids.

- *Gas chromatography.* This is the most powerful multicomponent method for volatile compounds. The resolution of GC is considerably increased by application of capillary columns, and its specifity is improved by coupling with a mass spectrometer (GC–MS) or other instruments, such as GC–IR, GC–NMR, GC–UV, and GC–ICP. The large potential information amounts of about 10^6 to 10^{10} bits resulting from such coupling techniques, particularly GC–HRMS, demanded an additional coupling with a computer from the beginning; therefore the first integration of computers took place in this analytical field.

- *Liquid chromatography.* When high pressure techniques and efficient columns are applied to *high performance liquid chromatography (HPLC)*, its power approaches that of GC. Recently the field of HPLC has been enlarged to include the analysis of inorganic components. *Ion chromatography* can be applied to multicomponent analysis of a large number of ions, cations as well as anions, and inorganic as well as organic. By coupling HPLC with MS, the efficiency can be much increased (*thermospray and electrospray mass spectrometry*; *TS–MS, ES–MS*), and by combination with atomic spectroscopic methods, such methods become serviceable for *speciation analysis* (HPLC–AAS, HPLC–ICP).

- *Thin-layer chromatography*. Recently this method has also been extended by new techniques such as scanners and photodiode arrays, the latter of which has proved its worth in other analytical fields, particularly HPLC.

In general, the efficiency, applicability, and the information amount can be much improved by the combination of different techniques, namely by *coupled methods* (so-called *hyphenated methods*). Some principles of hyphenated techniques are shown in Table 7.1. Coupling of methods (hyphenated techniques) may prove to be the most effective strategy for improving performance parameters, particularly those relating to specifity, detection power, reliability, and speed.

Table 7.1. Some Selected Principles of Hyphenated Methods

Concern of Coupling	Principles	Examples
Observation and adjustment of samples, location of inclusions for microanalysis	Microscopy, laser ablation, field ion microscope	Laser OES, LA–ICP, FIM–MS, AFM, (atom probe)
Transportation of sample	Flow analysis	FIA–AAS
Separation	Chromatography	GC–MS, –FTIR HPLC–AAS, TS–, ES–MS
Enrichment	Solid phase extraction	FIA–SPE–AAS
Generation of species	Laser ablation, electroanalysis	LA–AAS, –ICP ECA–MS
Detection	MS, MSD (mass specific detection)	GC–MS, MS–MS HPLC–MS

7.2 INFORMATION AMOUNT OF RESULTS

In multicomponent analysis the important factors include not only the signal positions and intensities but also their resolvability, which affects information gain. The relation between the multicomponent analysis system input and output is given by a matrix with one of the following two components.

1. In qualitative or identification analysis, the conditional probabilities $P(A_i|z_j)$.
2. In quantitative analysis, the sensitivities S_{ij} of determination of analyte concentrations x_i $(i = 1, 2, \ldots, n)$ by signals z_j $(j = 1, 2, \ldots, k)$; $k \geq n$.

We will denote the conditional probability matrix **P** and the sensitivity matrix **S**. The analytical signal in a multicomponent analysis is distorted by random noise—as in single-component analysis—as well as by overlap with adjacent signal(s). The a priori and a posteriori distributions are the same as in single-component analysis or, in the determination or detection of trace components, as in the individual cases of trace analysis. The truncated normal distribution is better suited to the region of $x_D \leq \mu \leq x_I$.

The uncertainty remaining after qualitative or identification multicomponent analysis is usually expressed as the entropy for the conditional probabilities. The uncertainty of an individual detection by measurement of signals in a certain position z_j is given by the so-called component entropy

$$H\big(P(A|z_j)\big) = -\sum_{i=1}^{n} P(A_i|z_j) \text{lb} \, P(A_i|z_j) \tag{7.6}$$

where $j = $ const. The amount of information obtained during multicomponent identification or qualitative analysis in which n components are sought simultaneously by measuring signals in $k \geq n$ positions can be expressed as

$$M = -\sum_{i=1}^{n_0} P(A_i) \text{lb} \, P(A_i) + \sum_{j=1}^{k}\sum_{i=1}^{n} P(A_i|z_j) \text{lb} \, P(A_i|z_j) \tag{7.7}$$

This amount of information is dependent on the selectivity of the procedure. We will show later how the selectivity of qualitative multicomponent analysis can be characterized by means of entropy that characterizes the a posteriori uncertainty from Eq. (7.7) and is given as

$$H(P(A|z)) = - \sum_{j=1}^{k} \sum_{i=1}^{n} P(A_i|z_j) \text{lb} \, P(A_i|z_j) \qquad (7.8)$$

It can be considered the sum of component entropies, by Eq. (7.6), over all $j = 1, \ldots, k$. It is calculated from the elements of the **P** matrix $(n \times k)$. If it is a unit diagonal matrix—that is, $P(A_i|z_i) = 1$ and $P(A_i|z_j) = 0$ for $i \neq j$—the procedure is perfectly selective and $H(P(A|z)) = 0$.

Associated with the selectivity of multicomponent qualitative or identification analysis is the so-called equivocation

$$E = - \sum_{j=1}^{k} \sum_{i=1}^{n} P(z_j) P(A_i|z_j) \text{lb} \, P(A_i|z_j)$$

$$= - \sum_{j=1}^{k} P(z_j) H(P(A|z_j)) \qquad (7.9)$$

where $H(P(A|z_j))$ is the entropy for the conditional probabilities according to Eq. (7.6). The probability that there appears a signal $P(z_j)$, which in Eq. (7.9) has the meaning of the "weight" attributed to the entropies, is

$$P(z_j) = P(A_i) \sum_{i=1}^{n} P(z_j|A_i) \qquad (7.10)$$

Since $P(A_i)$ is largely equal to $1/n$, equivocation is usually calculated as

$$E = - \frac{1}{n} \sum_{j=1}^{k} \sum_{i=1}^{n} P(z_j|A_i) P(A_i|z_j) \text{lb} \, P(A_i|z_j) \qquad (7.11)$$

Equivocation can be employed to characterize the selectivity of the procedure, to optimize the procedure of a qualitative or identification multicomponent analysis, to quantify the signal overlap, and so on.

These considerations rely on the assumption that the analyte concentration x_i is invariably much higher than the detection limit x_D, so minor changes in the former have no effect on the appearance of the signal, be it a color reaction in a spot test or the presence of a line at a given wavelength in the spectrum. If, however, an analyte whose concentration approaches the detection limit is to be detected, $P(z_j|x_i)$ depends on x_i. This dependence was studied by Liteanu and Rica (1979) who demonstrated that its shape corresponds to that of the normal distribution function. The probability $P(z_j|x_i)$, though, is codetermined by the sensitivity of the reaction or, in instrumental analysis, by the sensitivity of the detector.

In qualitative analysis there are only two alternatives, namely whether or not the signal is discernible from the background. The conditional probability that the signal of an analyte concentration x_i appears is $P(y = 1|x_i)$ and that it does not appear is $P(y = 0|x_i) = 1 - P(y = 1|x_i)$; therefore, Eckschlager and Král (1984) introduced the entropy

$$H\big(P(x_i|y = 1)\big) = - \sum_{i=1}^{n} P(x_i|y = 1)\,\mathrm{lb}\, P(x_i|y = 1) \quad (7.12)$$

with the equivocation

$$E = \sum_{y_j=0}^{1} P(y_j) \cdot H\big(P(x|y = 1)\big) \quad (7.13)$$

These quantities, however, are only reasonable for a rather narrow concentration region for which $0 < P(y = 1|x_i) \le 1$. This concentration interval is sometimes referred to as the *region of uncertainty* of the qualitative detection.

All of the above considerations refer to the case where each analyte has at least one signal at position z_j, as is true in chromatography, for example. In other cases, such as optical emission spectroscopy, each analyte has several signals in positions z_j, $j =$

$1, \ldots, m$. We can, more formally, introduce the quantity

$$H(P(A_i|z)) = - \sum_{j=1}^{k} P(A_i|z_j) \text{lb } P(A_i|z_j) \qquad \text{for } i = \text{const.}$$

(7.14)

which we will call the *signal entropy*. But more important than the value of this entropy is whether what interferes with the signals of analyte A_i in the various positions z_j, $j = 1, \ldots, k$, is always the same component or different components and whether these are major, minor, or trace components.

The uncertainty of the detection of analyte A_a is reduced considerably if the absence of a signal in another position gives evidence of the absence of a component A_i, $i \neq a$, whose presence interferes with the signal: Actually, by Eq. (7.7), we would then have $P(A_i) = 0$. It may be possible to find the corresponding mathematical expression for the information content of the signal, making allowance for whether, and to what extent, another signal interferes, though this would not be of practical use.

7.3   INFORMATION AMOUNT OF ANALYTICAL METHODS

In quantitative multicomponent analysis the maximum potential information amount M_p of a whole sequence of signals (spectrum, chromatogram, polarogram, etc.; see Fig. 7.1) can be determined as

$$M_p = \frac{z_{\max} - z_{\min}}{\Delta z} \text{lb} \frac{y_{\max} - y_{\min}}{\sigma_y \sqrt{(2\pi e)}}$$

(7.15)

where $z_{\min}$, $z_{\max}$, $y_{\min}$, and $y_{\max}$ are the lowest or highest values of the positions z and signal intensities y recorded by the instrument ($z_{\max} - z_{\min}$ recording range) and Δz is the *signal resolution* (i.e., the shortest distance required for the adjacent signals not to affect each other). These distances are different for different signal profiles and depend on several factors:

1. Whether a quantitative analysis or qualitative and identification analyses are carried out.

2. Whether the signal intensity is the peak height or peak area.
3. Whether the signals are evaluated univariate or multivariate.

Whereas in classical quantitative analysis the signal resolution should be approximately double the signal half-width $\Delta z_{1/2}$ (particularly in the case of evaluation of peak height; see Doerffel and Eckschlager 1981),

$$\Delta z \approx 2\Delta z_{1/2} \qquad (7.16)$$

in qualitative and identification analysis it is possible to recognize also signals that are closer together than as noted in Eq. (7.16). A distance of

$$\Delta z \approx \Delta z_{1/2} \qquad (7.17)$$

is frequently sufficient to detect adjacent signals and also to determine them quantitatively by means of multivariate techniques like PLS (partial least squares; see Sjöström, Wold, Lindberg, Persson, and Martens 1983).

On closer examination the maximum potential information amount, by Eq. (7.15), represents a special case where the resolution of signals is constant over the whole recording range. This is what several of the spectroscopic techniques show; consider the case of spectra obtained by use of grating spectrographs.

The potential information amount is characterized by the (potential) *analytical resolution power* N_p (Doerffel and Hildebrandt 1969):

$$M_p = N_p \text{ lb } m \qquad (7.18)$$

where m is the concentration resolution [see Eq. (7.2)]. In the most general case N_p is given (Danzer 1975) by

$$N_p = \int_{z_{min}}^{z_{max}} \frac{dz}{\Delta z} \qquad (7.19)$$

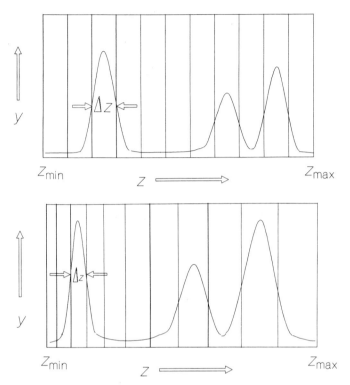

Figure 7.6. Analytical resolving power for Δz = const. (*top*) and $\Delta z = f(z)$ (*bottom*)

For Δz = const. (see Fig. 7.6, top) Eq. (7.19) yields

$$N_p = \frac{z_{max} - z_{min}}{\Delta z} \tag{7.20}$$

By Eq. (7.20), the potential information amount obtained by Eq. (7.18) becomes that of Eq. (7.15).

Frequently the signal half-width is a function of the recorded quantity z (see Fig. 7.6, bottom), for example, in the form $\Delta z^{-1} = f(z)$. The analytical resolving power according to Eq. (7.19) then takes the general form

$$N_p = \int_{z_{min}}^{z_{max}} f(z)\, dz \tag{7.21}$$

We obtain such a variable signal resolution, for example, with the signal values found for chromatographic and several spectroscopic methods. In the latter case the relation is expressed by the spectral resolution power $R(z) = z/\Delta z$ [e.g., $R(\lambda) = \lambda/\Delta\lambda)^1$] and Eq. (7.21) becomes

$$N_p = \int_{\lambda_{min}}^{\lambda_{max}} \frac{R(\lambda)}{\lambda} d\lambda \qquad (7.22)$$

If the resolution power R is practically constant, we have (Kaiser 1970; Danzer 1975):

$$N_p = R \ln \frac{\lambda_{max}}{\lambda_{min}} \qquad (7.23)$$

Equations (7.20) and (7.23) characterize special cases for constant signal resolution Δz and constant resolution power R, respectively. In practice, only rough approximations can be obtained in some cases, such as those involving for prism spectrographs. Where there is doubt, Eq. (7.21) can be used with a known function or an estimate of $f(z)$.

The results for a prism spectrograph Q 24 (Carl Zeiss, Jena) obtained in different ways are compared in Table 7.2. Clearly the approximations become worse the wider the integration limits are. Moreover we find confirmation of the fact that analytical resolving power, and therefore the information amount, strongly decrease with increasing wavelength.

The analytical resolving power determines directly the potential information amount M_p of analytical methods

$$M_p = N_p \cdot I_{max} = N_p \cdot \text{lb } m \qquad (7.24)$$

The maximum potential resolving power, and therefore the potential information amount, can take considerably high values. Estimates for such values (by order of magnitude) for some important methods are given in Table 7.3.

[1] A similar relation characterizes the mass resolution in mass spectrometry, $R(m) = m/\Delta m$.

Table 7.2. Analytical Resolving Power N of a Prism Spectrograph of Medium Dispersion

Integral Limits (Wavelength in nm)		Analytical Resolution Power N according to		
λ_{min}	λ_{max}	Eq. (7.21)	Eq. (7.23)	Eq. (7.20)
200	500	11,110	8,900	8,350
a: 200	300	6,670	6,340	6,250
b: 300	400	2,855	2,770	2,780
c: 400	500	1,585	1,570	1,570
$a + b + c$ 200	500	11,110	10,680	10,600

The information gain achieved by *component separation* (e.g., in chromatography or electrolysis) or by *signal resolution* (e.g., experimentally by a prism or grating in emission spectrometry or by computational deconvolution)—irrespective of whether identification, qualitative, or quantitative analysis is involved—can be determined for $i = 1, \ldots, n$ components A_i during the measurement of

Table 7.3. Analytical Resolving Power N_p and Potential Information Amount M_p of Selected Analytical Methods

Method	N_p	M_p/bit
Spot tests	1	1
Titration	10	100
Optical emission spectroscopy		
• by grating instruments	200,000	1,600,000
• by prism instruments	10,000	80,000
• by "quantometers"	60	600
Mass spectroscopy	500	4,000
• high-resolution MS	200,000	1,600,000
UV–Vis spectrophotometry	50	500
X-ray spectrometry		
• wavelength-dispersive	5,000	50,000
• energy-dispersive	500	4,000
Infrared spectrometry	1,000	8,000
Gas chromatography	1,000	8,000
• Capillary GC	10,000	80,000

signals $y(z_j)$ in position $j = 1, \ldots, k$ $(k \geq n, j = \text{const.})$ as

$$I_{\text{sep}}^{(j)} = \text{lb } n + \sum_{i=1}^{n} a_{ij} \text{ lb } a_{ij} \tag{7.25}$$

It invariably holds that $0 \leq I_{\text{sep}}^{(j)} \leq \text{lb } n$; the maximum value is attained at a perfect separation, the zero value, if no separation or signal resolution for the components has taken place. In the case of qualitative analysis we insert into Eq. (7.25)

$$a_{ij} = P(A_i|z_j) \qquad \text{for } j = \text{const.} \tag{7.26}$$

In the case of quantitative analysis, we add

$$a_{ij} = \frac{S_{ij}}{\sum_{i=1}^{n} S_{ij}} \qquad \text{for } j = \text{const.} \tag{7.27}$$

where a_{ij}, which lie within the interval of $\langle 0, 1 \rangle$, are elements of matrix $\mathbf{A} = (a_{ij})$ of dimension $n * k$ $(k \geq n)$. If a component cannot be detected or determined by a signal in position z_j, we have $a_{ij} = 0$ and set $-a_{ij} \text{ lb } a_{ij} = 0$.

If the signal intensity is measured in k positions, $j = 1, \ldots, k$ $(k \geq n)$, the amount of information obtained by separation is

$$M_{\text{sep}} = k \text{ lb } n + \sum_{j=1}^{k} \sum_{i=1}^{n} a_{ij} \text{ lb } a_{ij} \tag{7.28}$$

If each analyte is detected or determined by measuring a single signal, then $k = n$ and

$$M_{\text{sep}} = n \text{ lb } n + \sum_{j=1}^{n} \sum_{i=1}^{n} a_{ij} \text{ lb } a_{ij} \tag{7.29}$$

The average information gain per component, attained by separation of n components, is

$$I_{sep} = \frac{1}{n} M_{sep} = \text{lb } n + \frac{1}{n} \sum_{j=1}^{n} \sum_{i=1}^{n} a_{ij} \text{ lb } a_{ij} \qquad (7.30)$$

so that invariably $0 \le I_{sep} \le \text{lb } n$. For perfectly isolated signals, $I_{sep} = \text{lb } n$, whereas if no separation occurs, $I_{sep} = 0$. Further details can be found in Eckschlager (1989).

7.4   SELECTIVITY

Selectivity is a very important property in multicomponent analysis. It was first defined by Kaiser in 1972 [see Eq. (4.3)]. Nowadays there exist at least eight different definitions, but none is generally valid.

In 1989 Eckschlager demonstrated that selectivity should be understood as a continuous property. But it is convenient to discriminate between cases where selectivity is high enough not to have an effect on the accuracy of results and cases where the nonselectivity is so appreciable that signal overlap must be taken into account in signal processing.

The general applicability characteristics of selectivity is satisfied by a computation of a parameter similar to entropy. For the measurement of a signal in position z_j, this quantity is

$$H(a_j) = - \sum_{j=1}^{n} a_{ij} \log a_{ij}$$

$$\sum_{j=1}^{n} a_{ij} = 1 \qquad \text{for } j = \text{const.} \qquad (7.31)$$

whereas for the whole analytical procedure, it is

$$H(a_{ij}) = - \sum_{j=1}^{k} \sum_{i=1}^{n} a_{ij} \log a_{ij} \qquad (7.32)$$

Crucial for obtaining this value by Eq. (7.32), which is zero for a perfectly selective procedure, is the matrix $\mathbf{A} = (a_{ij})$. This matrix is

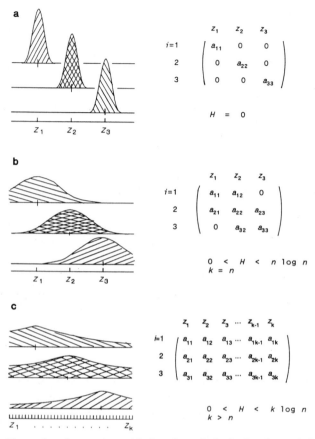

Figure 7.7. Entropies for various kinds of analytical signals and for various matrices $\mathbf{A} = (a_{ij})$

square for $n = k$. For the diagonal matrix, where $a_{ii} > 0, a_{ij} = 0$ for $i \neq j$ (the whole analytical process is perfectly selective), we have $H(a_{ij}) = 0$; otherwise, $0 < H(a_{ij}) \leq k \log n$ ($n \leq k$) (see Fig. 7.7).

It is clear that improvement of the resolving power of the instrument, as well as better signal resolutions, computational deconvolution or transformation, cross-correlation, and so on, will reduce uncertainty and thus increase the amount of information that can be obtained by multicomponent analysis. Mathematical methods of resolution improvement are frequently accompanied by an increase of noise (i.e., random error). As the standard deviation σ_y increases, so does the a posteriori uncertainty of the individual determinations. The total amount of information obtained by multicomponent analysis, however, usually increases because of the improvement in resolution.

In nonselective procedures there is frequently the overlap of sequences of signals to consider; that is, overlapping signals must be "separated" numerically. Signal-processing methods like deconvolution are designed for this purpose, or modeling techniques such as PLS for finding latent variables can be used.

The quantitative analysis of selectivity by Eq. (7.32) in terms of Eqs. (7.26) and (7.27) is generally also valid for qualitative analysis and identification. In the quantitative analysis where strong overdetermined systems of equations are involved, this equations (with $k > n$) takes the form

$$y(z_j) = \sum_{i=1}^{n} S_{ij} \cdot x_i \qquad (7.33)$$

with n unknowns, which are solved for the condition of the least sum of squares of differences, $\Sigma(x_{ij} - \bar{x}_i)^2 = \min$. This minimization may be carried out with the original variables (classical techniques) as well as latent variables (multivariate techniques) by principal component regression (PCR) or partial least squares (PLS) modeling.

By Eq. (7.32) we see that the characteristics of selectivity are based on the sensitivity matrix. This matrix is also appropriate for some other important selectivity concepts, namely that of Kaiser (1972), given by Eqs. (4.1) and (4.3), and that of Kowalski (see Sharaf, Illman, and Kowalski 1986). The latter are related to matrix notation of multicomponent analysis:

$$\mathbf{Y}_{s,k} = \mathbf{X}_{s,n}\mathbf{A}_{n,k} \qquad (7.34)$$

where s is the number of standard samples used for calibration, k the number of sensors, wavelengths (generally different elements of responses) used, and n the number of analytes of interest. In the case of a square matrix ($k = n$), sensitivity is given by the absolute value of the determinant of **A**

$$S = |\det \mathbf{A}| \tag{7.35}$$

In the more general case where $k > n$, the sensitivity is defined by

$$S = \sqrt{\det(\mathbf{A}^T\mathbf{A})} \tag{7.36}$$

The greater S is by Eq. (7.35) or (7.36), the better is the sensitivity and the more precise the determination of the n analytes. This general concept of selectivity has been proved useful for quantitative multicomponent analyses in spectralphotometry and chromatography (Otto and Wegscheider 1986).

7.5   RELEVANCE AND REDUNDANCY

Beyond its factual content any information requires interpretation. The two properties of content and meaning are important in assessing the relevance of results in solving analytical problems.

In multicomponent analysis the kind and number of components that can be determined simultaneously are given by the analytical method. The greater the number of components, the larger is the analytical resolving power. The information obtained can have different relevance for different analytical problems. Therefore we introduce the notion of *usable amount of information*,

$$M_E = \sum_{j=1}^{n} I_i \cdot k_i \tag{7.37}$$

where the information relevance coefficient k_i, for component A_i in the given problem is $k_i \in \langle 0, 1 \rangle$. If several analytical methods, $j = 1, \ldots, q$ $(q \geq 2)$, have to be used to gain the information, the usable amount of information obtained by combining q methods is

$$M_E = \sum_{i=1}^{n} \sum_{j=1}^{q} I_{ij} \cdot k_{ij} \tag{7.38}$$

where I_{ij} is information gain obtained by the jth method on the ith analyte and k_{ij} is the corresponding coefficient of relevance, $k_{ij} \in \langle 0, 1 \rangle$.

It is clear that we always have $0 < M_E \le M$ irrespective of whether the results have been obtained by a single multicomponent method or by a combination of several methods of single- or multicomponent analysis. Details concerning the amount of information obtained by combining several methods are given by Eckschlager and Stepánek (1979, ch. 6), who distinguish between analytical methods combined in series (successive) and in parallel (simultaneous).

The coefficient of relevance for the ith analyte can be regarded as either a constant (i.e., the static model where $k_i = $ const.) or a variable that varies with the information content of the result, $\mathrm{d}k_i/\mathrm{d}I_i = f(I_i)$ (i.e., the dynamic model which was introduced by Eckschlager and Stepánek 1985, 1987). Two kinds of dynamic models have found application. The first

$$k_i = \begin{cases} 0 & I < I_N \\ k_{i,\max}\{1 - \exp[-a(I - I_N)]\} & I \ge I_N \end{cases} \qquad (7.39)$$

is used where the information is only relevant if its content I is equal to or higher than the required content I_N ($I \ge I_N$). The dependence of k_i on I is shown in Fig. 7.8a.

The second model

$$k_i = k_{i,\max}\{1 + \exp[-a(I - I_{1/2})]\}^{-1} \qquad (7.40)$$

is suitable if we require that $I_{\min} \le I \le I_{\max}$; at $I = I_{1/2}$ we have $k_i = \frac{1}{2}k_{i,\max}$. Equation (7.40) is actually Robertson's growth rule. The dependence of k_i on the information content is shown in Fig. 7.8b.

The static model is actually a special case of the dynamic model for $dk_i/dI_i = 0$, or $I \gg I_N$ in Eq. (7.39) and $I \gg I_{\max}$ in Eq. (7.40). The k_i value in the static model, and the $k_{i,\max}$ values in the dynamic model, must be determined in advance for each analyte.

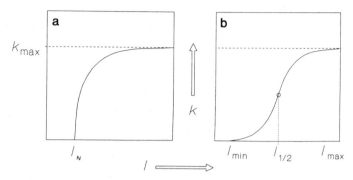

Figure 7.8. Dependence of the coefficient of relevance on the information content in Eqs. (7.39) and (7.40)

There are four possible choices:

k_i = 1.0 for highly relevant analytes (indispensable for a given problem)

k_i = 0.75 for relevant analytes

k_i = 0.5 for less relevant analytes

k_i = 0.25 for analytes not actually but potentially relevant

Of course some individual subjectivity is necessary, since the values involved are fuzzy. In mathematics the treatment of such indefinitely determined quantities and facts has been termed *fuzzy set theory* (Zadeh 1968), and recently this idea has been also used in analytical chemistry (Blaffert 1984; Otto and Bandemer 1986).

It is useful to regard the coefficient of relevance as a value of the function of membership of information on ith analyte in the fuzzy subset of information relevant to the solution of the problem in question. This enables us to solve the case of mutually dependent parts of information. For two equally relevant but factually quite independent parts of information with contents I_1 and I_2, for example, we have

$$M_E = I_1 k_1 + I_2 k_2 = k(I_1 + I_2) = k \sum_{i=1}^{2} I_i \qquad (7.41)$$

where $k = k_1 = k_2$. If, however, information about one component also contains some information about another component, Eq. (7.41) holds true, but $k \leq k_1 = k_2$. If, on the contrary, information about component A_1 and information about component A_2 do not separately enable a decision but together contribute appreciably to a decision, then $k \geq k_1 = k_2$.

The irrelevance of some determined component content X_i may be due to information content that is lower than necessary, or is at $k_i = 0$. The case $I(r; p, p_0) \leq 0$ can occur if the condition $I(r; p, p_0)_0 > \frac{1}{2}z(\alpha)^2$ is not met (see Section 6.6.2). If a component can be determined by more than one method, we choose that method for which I_{ij} is highest. If we want to determine some analyte content X_i by more than one method (e.g., by two methods) and the difference between the results is not statistically significant (e.g., at the level $1 - \alpha = 0.95$), we speak about statistical convergence of results of the different methods. If the two methods differ in their chemical or physicochemical nature, such convergence is a strong proof of the likelihood of the results. In practice, the convergence of results of different methods is tested by interlaboratory comparisons (see Chapter 9).

Another important property of experimentally acquired information is its redundancy, which, when defined generally, is called *absolute redundancy*

$$R = M_{max} - M \tag{7.42}$$

where M_{max} is the maximum attainable information amount and M the actual information amount needed for solving a certain problem. For single-component analysis the redundancy is given by

$$R = I_{max} - I \tag{7.43}$$

which corresponds to Eq. (7.42). Frequently, relative redundancy is used instead of absolute redundancy. For the relative redundancy in single-component analysis, we have

$$r = 1 - \frac{I}{I_{max}} = \frac{I_{max} - I}{I_{max}} = \frac{I_L}{I_{max}} \tag{7.44}$$

and for the relative redundancy in multicomponent analysis,

$$r = 1 - \frac{M}{M_{max}} = \frac{M_{max} - M}{M_{max}} = \frac{M_L}{M_{max}} \qquad (7.45)$$

Here I and M are the information gain and amount of information, respectively; the subscript "max" refers to the maximum attainable value and the subscript "L" to the lost or "excessive" information gain or amount. We can also write

$$I_L = r \cdot I_{max} \qquad (7.46)$$

It is clear that $0 \le r \le 1$. For cases where repetitions must be performed, relative redundancy can be expressed by rewriting Eq. (7.44) as

$$r = 1 - \frac{I_n}{nI_1} = \frac{nI_1 - I_n}{nI_1} \qquad (7.44a)$$

where I_n is the information gain attained by performing n repeated determinations.

Clearly a large number of repeated results give more redundant than useful information. But some redundancy is present even if a single determination is performed, since an analysis check of a sample is customarily carried out after several analyses of its contents. The existence of the a posteriori uncertainty simply does not enable reliable information to be obtained free of some redundancy. With this protection, for example, against an outlying value, we also have more confidence in the results if they are precise as well as adequately redundant.

In this connection the term "promoting" redundancy is used. In contrast to "blank" redundancy, the notion of promoting redundancy both supports and benefits information gain. We can illustrate this by an example from optical emission spectrography. For about 80 elements there exist hundreds of thousands of spectral lines, but the spectrochemists focus on relatively few of them: the so-called main detection lines according to Gerlach et al. (1930, 1933, 1936). The main detection lines are a set of lines associated

**Table 7.4. Information Gain (in Bits) from Signals Evaluated
for Different a priori, $P(z_j)_0$, and a posteriori Probabilities, $P(z_j)$**

$P(z_j)_0$	$P(z_j)$				
	0.90	0.95	0.99	0.999	1.00
0.50	0.53	0.72	0.92	0.99	1.00
0.90		0.19	0.39	0.46	0.47
0.95			0.20	0.27	0.28
0.99				0.07	0.08

with promoting redundancy, whereas the other lines fall into the category of blank redundancy. The number of lines that enables a relative increase of information is rather low. The identification of only one spectral line does not suggest with certainty an analyste's presence, rather it only increases the probability of its presence.

Table 7.4 gives a stepwise sequence of three signals for the same analyte. For the probability sequence of 0.50 (a priori), 0.90, 0.95, and 0.99, the respective information gain is 0.53 bit, 0.19 bit, and 0.20 bit, with a total 0.92 bit. Looking for additional lines would increase the information gain only by 0.08 bit. Likewise, as the probability increases by the successive identification of lines, the nature of redundancy changes from promoting to blank redundancy.

If the information gain I_N or the amount of information M_N required to solve a particular problem is known, then

$$r = \begin{cases} \dfrac{I - I_N}{I} & I > I_N \\ 0 & I \leq I_N \end{cases} \tag{7.47}$$

and

$$r = \begin{cases} \dfrac{M - M_N}{M} & M > M_N \\ 0 & M \leq M_N \end{cases} \tag{7.48}$$

At $I < I_N$ or $M < M_N$, the results are barely relevant or irrelevant. Therefore it is expedient always to determine the relevance and redundancy of the results simultaneously. Too high a redundancy indicates that the potential of the analytical method employed in the multicomponent analysis is utilized inefficiently and also uneconomically.

In instrumental multicomponent analysis, redundancy is frequently caused by the high stock of signals, expressed by the large potential resolving power. In practice, we need ideally n signals to analyze n components; in consideration of promoting relevancy a set of about $3n$ may be useful in some cases. The surplus of $N_p - n$ is decisive for redundancy in multicomponent analysis. By Eqs. (7.42) and (7.18) we have

$$R = (N_p - n)\operatorname{lb} m \qquad (7.49)$$

or, expressed in terms of relative redundancy,

$$r = 1 - \frac{n}{N_p} \qquad (7.50)$$

By Eqs. (7.49) and (7.50), $R_{\text{prom}} = 2n \operatorname{lb} m$ or $r_{\text{prom}} = 2n/N_p$ for promoting redundancy, and therefore $R_{b1} = (N_p - 3n)\operatorname{lb} m$ or $r_{b1} = 1 - 3n/N_p$ for blank redundancy. Information that has been found to be experimentally redundant can be obtained as well by deduction or from the literature on it. Redundancy and relevance have been dealt with in detail by Eckschlager and Stepánek (1987).

7.6 PRACTICAL APPLICATIONS

The effect of redundancy on relevant information in multicomponent analysis will be considered in an example of trace analysis by atomic emission spectrography: In highly purified alumina used in the preparation of laser rods, impurities, especially transition elements, can affect the optical properties of the laser rods. Therefore, as many elements as possible have to be analyzed (Schrön, Krieg, and Danzer 1992), namely Fe, Mn, Cr, Cu, Ni, Co, V, Ti, Ca, K, Na, Si, Mg, Cl, S, and P whose detection powers are given in Table 7.5.

Table 7.5. Analytical Parameters of Analysis of High-Purity Alumina by Arc Optical Emission Spectrography

Analyte	k	x_2/ppm	x_D/ppm	s_x/ppm	I_+/bit	I_-/bit
Fe	1.0	100	2	1	5.33	5.64
Mn	1.0	20	0.5	0.2	5.32	5.32
Cr	1.0	20	0.2	0.2	5.32	6.64
Cu	1.0	20	0.2	0.2	5.32	6.64
Ni	1.0	20	0.5	0.2	5.32	5.32
Co	1.0	10	0.5	0.1	4.32	4.32
V	1.0	10	0.5	0.1	4.32	4.32
Ti	1.0	100	0.5	0.2	5.33	7.64
Cl[2]	0.75	1000	20	10	6.33	5.64
S[2]	0.75	100	5	1	5.33	4.32
Ca[2]	0.5	1000	5	10	5.33	7.64
K[2]	0.5	200	5	1	5.33	5.32
Na[2]	0.5	1000	5	10	5.33	7.64
P[2]	0.5	1000	10	10	5.33	6.64
Si	0.25	2000	5	20	5.33	8.64
Mg	0.25	1000	1	10	5.33	9.97

Note: k relevance coefficients, x_2 upper expectation limit, x_D detection limit, s_x standard deviation estimated from a relative standard deviation $s_{x,rel} = 0.1\%$; $I_- = I(p, p_0)_- = $ lb(x_2/x_D) information gain by Eq. (6.31), $I_+ = I(p, p_0)_+ = $ lb$(x_2/\Delta x) = $ lb$[x_2\sqrt{n}/[s_x t(a, f)]] \approx$ lb$\{x_2/[\sigma\sqrt{2\pi e}]\}$ information gain approximately by Eq. (6.36).

We will consider two special cases:

1. All impurities are found in the following contents (in ppm): Si: 100; Ca, Na, Mg, P: 50; Cl: 25; K: 10; Fe, Ti, S: 5; Mn, Cr, Cu, Ni, Co, and V: 1;[2] the information amount then is $M_{(a)} = \Sigma k_i I_{i+} = 62.65$ bit;

2. No impurities are found; then the information amount is $M_{(b)} = \Sigma k_i I_{i-} = 64.86$ bit.

Since the values of the detection limits x_D are similar to the values of the absolute standard deviations s_x, the information gains

[2]Cl, S, and P were determined by spark source mass spectrography at the Institute of Solid State Physics and Materials Science, Dresden, by Stahlberg (1988); Na, K, and Ca were determined by flame spectrometry; all the other elements by arc OES (Schrön, Krieg, and Danzer 1991).

are comparable in both the positive and negative results. It is no accident that the negative results of trace analyses are characterized by higher information gain than the positive results; in our example (Table 7.5) it applies to 9 of the 16 elements.

Besides the relevant information amount $M_{(a)}$ and $M_{(b)}$, to some extent promoting redundancy is present in the evaluation of photographically recorded spectra, since more lines are considered than the main lines in deciding on case 1 or 2 and in obtaining the values in Table 7.5. If we assume that in the positive as well as in the negative case *one* additional signal is observed for the certainty of the decision to increase from 0.90 to 0.99 (see Table 7.4), then on average the information gain increases by 0.39 for each element. For the relevance coefficients that we have in our example, the amount of promotive-redundant information by which the information amounts $M_{(a)}$ and $M_{(b)}$ increase is

$$M_{p.r.} = (8 + 2 \times 0.75 + 4 \times 0.5 + 2 \times 0.25)0.39 = 4.68 \text{ bit}$$

If all the impurities are found, the real information amount is then

$$M_+ = M_{(a)} + M_{p.r.} = 67.33 \text{ bit}$$

and if no impurities are found, it is

$$M_- = M_{(b)+Mp.r.} = 69.54 \text{ bit}$$

Let us now categorize the different kinds of relevant and redundant information:

1. $M_{(a/b)}$, the amount of relevant information according to Eq. (7.37) based on the evaluation of one line corresponding to a certainty of about $P(z_j) = 0.90$.
2. $M_{+/-} = M_{(a/b)} + M_{p.r.}$, the sum of relevant plus promotive-redundant information based on more than one line corresponding to a certainty of about $P(z_j) = 0.99$.
3. Redundant information present in all lines belonging to analytes considered in the preceding two categories, $M_{r.a.}$

4. Other redundant information present in all lines of elements that do not affect the outcome of the analytical problem (unimportant components), $M_{r.u.}$

5. Regions in the spectra where characteristic signals do not appear (background regions) or cannot be detected because of wide-band interferences (e.g., the region of dicyane band spectrum), $M_{r.b.}$

6. Irrelevant redundancy, which is the difference between the maximum potential information amount M_p, by Eq. (7.15) characterizing the maximum signals possible, and the partial amounts under categories 1 through 5.

The boundaries between these categories of relevant and redundant information are sometimes blurred, and they can occasionally overlap. The history of atomic emission spectrography has showed that elements can turn out to be relevant only after analysis, such as in case of the spectrochemical analyses of highly purified substances used for determination of atomic weights (Ruthardt 1931; Gerlach and Riedl 1934).

The increase of information depends not only on decreased uncertainty as stated in categories 2 and 3 but also on the correlation between signals. For instance, van Marlen and Dijkstra (1976) found that the increase of information amount ΔM_k, resulting from the evaluation of the kth signal, can be described by

$$\Delta M_k = I_{k,u}\left(1 - \frac{\sum_{i=1}^{k-1}|r_{ik}|}{k-1}\right) \tag{7.51}$$

where $I_{k,u}$ is the information content of the kth, uncorrelated signal and r_{ik} the correlation coefficient between the kth and ith signals. The more the signals are correlated $(\sum|r_{ik}| \rightarrow k-1)$, the more ΔM_k will decline. This phenomenon was proved by van Marlen and Dijkstra (1976) for mass spectroscopy (see Fig. 7.9) and by Dupuis and Dijkstra (1975) for gas chromatography (see Fig. 7.10).

As Fig. 7.9 shows, in practice, when correlation is used, the information amount of mass-spectroscopic results increase less than

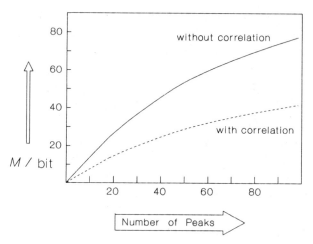

Figure 7.9. Information amount of mass spectra according to number of peaks (from Marlen and Dijkstra 1976)

is theoretically expected. An analogous effect can be seen in Fig. 7.10 for gas-chromatographic identification. The uncertainty in an analysis of an unknown mixture of components can be considerably decreased, and therefore the information amount increased, if the sample is investigated using different stationary phases. But, in practice, correlations between the results of different column materials always exist because of similarities in the interaction mechanisms. Therefore the theoretical increase of information amount cannot be obtained in practical investigations.

A special kind of redundancy is present in the mass-spectrometric elucidation of organic compounds: There exist ranges in the spectra in which characteristic key fragments cannot be expected. Furthermore there are restrictive relationships between certain peaks given by irregular mass differences, such as 4 to 14, 20 to 25, and (for chlorine-free compounds) 34 to 38 mass units.

The analytical strategies of relevancy and redundancy and the powerful methods of multicomponent analysis (which can analyze a large number of components in one execution) can only be evaluated with the aid information theory. Additional practical examples will be discussed in Chapters 8 and 9.

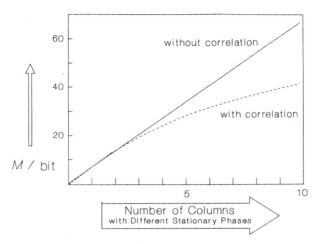

Figure 7.10. Information amount of GC retention index according to number of stationary phases (Dupuis and Dijkstra 1975)

REFERENCES

Blaffert, T. 1984. *Anal. Chim. Acta* **161**, 135.

Brown, S. D. 1986. *Anal. Chim. Acta* **181**, 1.

Danzer, K. 1973a. *Z. Chem.* **13**, 69.

Danzer, K. 1973b. *Z. Chem.* **13**, 229.

Danzer, K. 1975. *Z. Chem.* **15**, 158.

Danzer, K., Eckschlager, K., and Matherny, M. 1989. *Fresenius Z. Anal. Chem.* **334**, 1.

Danzer, K., Than, E., Molch, D., and Küchler, L. 1987. *Analytik—Systematischer Überblick*. Wissenschaftliche Verlagsgesellschaft, Stuttgart.

Doerffel, K., and Hildebrandt, W. 1969. *Wiss. Z. Techn. Hochsch. Leuna-Merseburg* **11**, 30.

Dupuis, F., and Dijkstra, A. 1975. *Anal. Chem.* **47**, 379.

Eckschlager, K. 1991. *Collect. Czech. Chem. Commun.* **56**, 506.

Eckschlager, K., and Stepánek, V. 1979. *Information Theory as Applied to Chemical Analysis*. Wiley, New York.

Eckschlager, K., and Stepánek, V. 1985. *Analytical Measurement and Information*. Research Studies Press, Letchworth.

Eckschlager, K., and Stepánek, V. 1987. *Chemometrics Intell. Lab. Syst.* **1**, 273.

Eckschlager, K., Stepánek, V., and Danzer, K. 1990. *J. Chemometrics* **4**, 195.

Gerlach, W., and Schweitzer, E. 1930. *Die chemische Emissionsspektralanalyse*, I. Teil. Grundlagen und Methoden. Voss, Leipzig.

Gerlach, Wa., and Gerlach, We. 1936. *Die chemische Emissionsspektralanalyse*, II. Teil. Anwendungen in Medizin, Chemie und Mineralogie. Voss, Leipzig.

Gerlach, W., and Riedl, E., 1934. *Z. anorg. allg. Chem.* **221**, 102.

Gerlach, W., and Riedl, E. 1939. *Die chemische Emissionsspektralanalyse*, III. Teil. Tabellen zur quantitativen Analyse. Voss, Leipzig.

Kaiser, H. 1970. *Anal. Chem.* **42** (2) 24A; (4) 26A.

Kaiser, H. 1972. *Fresenius Z. Anal. Chem.* **260**, 252.

Lindberg, W., Persson, J. A., and Wold, S. 1983. *Anal. Chem.* **55**, 643.

Marlen, G. van, and Dijkstra, A. 1976. *Anal. Chem.* **48**, 595.

Otto, M., and Bandemer, H. 1986. *Chemometrics Intell. Lab. Syst.* **1**, 71.

Otto, M. and Wegscheider, W. 1986. *Analyt. Chim. Acta* **180**, 445.

Ruthardt, K. 1931. *Z. anorg. allg. Chem.* **195**, 15.

Schrön, W., Krieg, M., Wienke, D., Wagner, M., and Danzer, K. 1992. *Spectrochim. Acta* **47B**, 189.

Sharaf, M. A., Illman, D. L., and Kowalski, B. R. 1986. *Chemometrics*. Wiley, New York.

Sjöström, M., Wold, S. Lindberg, W. Persson, J. A., Martens, H. 1983. *Anal. Chim Acta* **150**, 61.

Stahlberg, U. 1988. Unpublished report.

Zadeh, L. A. 1968. *Inf. Control* **8**, 338.

CHAPTER
8

OPTIMUM ANALYTICAL STRATEGIES

Over the past decades rapid progress occurred in the development of new diverse chemical and physical analytical methods, and measuring instrumentation and computer techniques were improved such that analytical problems that were altogether intractable until recently can be solved now. But new qualitative problems have emerged.

Today the quantities measured are not as clearly defined as mass, volume, and concentration were in the past. For instance, the scatter of measurement results is no longer negligible for the value measured, calibration poses various metrological problems, and blank must frequently be subtracted. In choosing the method and optimum working conditions, one has to take into account many factors that often do not manifest themselves in a clear-cut manner and in mutual interactions.

As in other branches of science, narrow specialization can confine the reasoning and working habits of a chemical analyst who might frequently prefer his or her "proven" method to a method that is actually optimum for a certain task. Economic aspects are often considerations as well.

Today chemical analysts seeking an optimum analytical strategy have had to increasingly turn to system- and information-theoretic approaches. A basis for the information-theoretic approach as it relates to the optimization of experimentally obtained information about chemical composition, was laid in the early 1980s by Doerffel and Eckschlager (1981; upgraded in cooperation with Henrion 1990).

Finding the optimum analytical or chemometrical strategy usually proceeds by the following steps:

1. Choose the most suitable analytical system, with particular attention to the suitability of its analytical method or technique.

2. Optimize the analytical procedure by experimental design and statistical analysis.
3. Double-check computerized signal processing by subjecting the measured signals to filtering, auto- and cross-correlation, and accumulation techniques.
4. Extract useful chemical information from measured data by means of chemometrics, for data analysis and interpretation.

The optimum analytical strategy usually consists in selecting among sets of methods that are available in the laboratory or that the analyst is familiar with. In the ideal case this would be the immense set of all existing analytical methods applicable to the problem in question. The criterion for arriving at the choice might be the quantity it yields such as information gain, or number of quantities for multicriterial selection. For computations, however, "a priori" values of the parameters are available—with a limited precision—from the literature, manufacturer's printed materials or from an analyst's previous experience.

In the optimization process, an objective function is established for some simple quantity such as the signal intensity or the signal-to-noise ratio. Then an "a posteriori" value—the value actually attained under the existing conditions—is found by parameter estimate. This is the idea behind discriminating between the "a priori" information gain $I(r; p, p_0)$ which is a function of some parameter (e.g., σ) and the "a posteriori" gain $\hat{I}(r; p, p_0)$ which is determined from an estimate of that parameter (e.g., from $s = \hat{\sigma}$).

8.1   CRITERIA FOR CHOICE OF ANALYTICAL METHOD

In selecting the analytical method it is expedient to take into account more factors simultaneously or to choose a criterion that makes allowance for more properties of the method. This can be facilitated by using appropriate mathematical methods tailored to the specific features of the chemical analytics. To make clear all aspects that will govern the selection, we can assign weights to them, or we can choose a single criterion that includes the various parameters to the appropriate extent. In any case, whether or not we make a multicriterial selection using more criteria or using a single "complex" criterion, the following requirements must be taken into account:

1. The region of applicability of the method, that is, the region of sufficient precision, must cover the entire concentration range of $\langle x_1, x_2 \rangle$ for which the method will be used.
2. The results must be timely; that is, the duration of the analysis must not be longer than the time in which we objectively want to know the result after we have submitted the sample for analysis.
3. The results must possess adequate parameter values (precision, accuracy, selectivity, limits of detection and determination, etc.) with respect to samples of the composition and the purpose of the analysis.
4. The choice of the method should include the possibility of metrological backup and traceability for quality assurance of the results.
5. The whole process must be economically advantageous or at least tolerable.

Maximum requirements are unrealistic: One must always pick out those parameters that are indispensable for the problem solving and bring them to fulfillment. The selected parameters must be useful; any other parameters, which of course are of minor importance, are disregarded.

In multicriterial decision making, the alternative analytical methods, procedures, modifications, and so on, are arranged in the order of their value:

$$D = \sum_{i=1}^{n} d_i w_i \qquad (8.1)$$

where $i = 1, \ldots, n$ is the number of partial criteria with values d_i transformed into a suitable interval (e.g., $\langle 0, 1 \rangle$ or 0 to 100%) and w_i are weights transformed into the interval $\langle 0, 1 \rangle$, too.

The criteria d_i must be so chosen that in the preferred case they all take their maximum values (if D is maximized) or minimum values (if D is minimized). In this arrangement some of the criteria may oppose one another, others may be correlated, and so on; thus the assignment of weights to criteria requires a suitable assignment strategy. It can be a *targeted strategy* where each criterion is assigned a relative preference, a *compromise strategy*, or an *"indifferent"* strategy where all criteria have the same weight. In the targeted strategy the weights are assigned not only according to the preferred properties but also according to which properties may be disregarded should a situation occur where a decision between several variants must be taken. An algorithm is composed for the entire procedure so that the multicriterial decision can be made by a computer with suitable software, such as ELECTRE (Roy 1987). Analytical applications of ELECTRE are given by Hálová (1989) and Pytela, Hálová, and Ludwig (1990).

In recent years expert systems have been used in the multicriterial selection of analytical methods, procedures, techniques, and variants. A general discussion of expert systems and their selection capabilities, from analytical methods to the organization of laboratories (LABGEN), can be found in papers of Klaessens and Kateman (1987) and Klaessens et al. (1989a, b). Also recent papers on the various applications of expert systems have dealt with method development and validation in HPLC (Mulholland et al. 1991), HPLC–UV (Gerritsen et al. 1992), method searching, selection, and conception for routine slag analysis, and Karl-Fischer titration (HELGA; Wünsch and Gansen 1989), design of procedures in ion chromatography (Wünsch and Schumnig 1991), in voltammetric trace analysis (Esteban et al. 1992), and in AAS (Lahari and Stillman 1992).

When a decision must be based on a single criterion, the chosen quantity should encompass more properties of the analytical method. In choosing the analytical method, this can be an information-theoretic performance parameter. Such quantities of general importance are drawn up in Table 8.1. Most of the quantities given in the table are used only in exceptional cases in analytical chemistry, such as for designating, recording or storage media and dynamic properties of instruments.

Quantities of more general importance are the *information performance IP* and *information profitability P*, which has been defined in various ways (see, e.g., Danzer and Eckschlager 1978; Doerffel

Table 8.1. Information-Theoretic Performance Parameters

Designation	Quantity	Unit	Application in Analytical Process
Information density (storage density, storage capacity)	$ID_a = M_p/a$ $ID_v = M_p/V$	bit/cm^2 bit/cm^3	Static detectors, e.g., photographic plate
Information capacity (channel capacity)	$IC = M_p/t$	bit/s, bit/h	Dynamic detectors, e.g., photomultiplier
Information flow (momentary information performance)	$IF = dM/dt$	bit/s, bit/h	Load of spectrometer channels, chromatography
Information performance	$IP = M/t_a$	bit/s, bit/h	Information power of analytical procedures

and Eckschlager 1981). In practice, the quantity is written

$$P = \frac{E \cdot M_E}{C} \exp\left(-\frac{t_a}{t_n}\right) \tag{8.2}$$

Here M_E is the usable amount of information according to Eq. (7.37) or (7.38), C is the cost of an analysis, and E is the efficiency coefficient which is usually defined as

$$E = e_1 \times e_2 \times \cdots \times e_k = \prod_{i=1}^{k} e_i \tag{8.3}$$

The partial coefficient e_i is defined as the ratio of the required value to the actual value of a property's characteristic (see, e.g., Danzer, Eckschlager, and Matherny 1989) $0 \leq e_i \leq 1$. The quantities t_a and t_n are, respectively, the duration of the analysis and the time needed for the results to become available.

In determining the costs, we must discriminate between fixed costs C_0 and variable costs C_1 which are dependent, for example, on the number of repeated results; we have $C = C_0 + C_1$. Costs C_1 can also vary according to the demands of the customer who has ordered the analysis; the customer must pay more for a more reliable analysis. Information profitability is expressed in bits per currency unit, c.u. (e.g., US$).

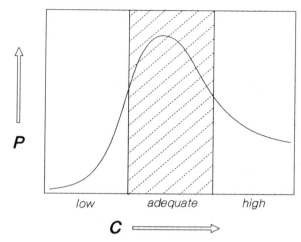

Figure 8.1. Information profitability and its relation to costs of analysis (*shaded area*)

The cost effectiveness of information depends on the time and money spent in solving a particular problem experimentally. This dependence is simplified in Fig. 8.1. As the figure shows, low cost yields few relevant results and excessive cost redundant results (more details on this relationship can be found in Doerffel, Eckschlager and Henrion 1990); the influence of cost on relevance in activation analysis was investigated by Obrusnik and Eckschlager 1993).

Figure 8.2 shows that the cost of an analysis usually grows more rapidly than the information gain of results (*top*). In the analyses involving the measurement of activity, a relative value of σ is inversely proportional to the square root of the cost of measurement. However, the relevance of information also affects the value of information in terms of cost; see Eqs. (8.2) and (7.37). In the middle and lower panels of Fig. 8.2, the P curve shows a maximum while the curve of information gain does not (for unbiased results). For the dynamic model of Eq. (7.40), to find the relevance coefficients, we assume that σ is between 1% and 10%. Then it can be seen that the cost of analysis grows too rapidly for highly precise results (low σ) to be cost effective.

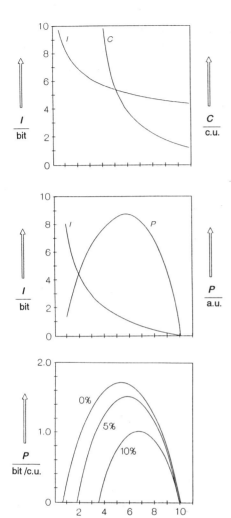

Figure 8.2. Dependence of information gain and cost on σ (*top*); information gain according to the dynamic relevance coefficient model and information profitability (*middle*) on σ; with parameters $x_2 - x_1 = 1000$ ppm, $x = 100$ ppm, $k = 1$; dependence of information profitability on σ (*bottom*) for three levels of bias δ (0%, 5%, 10%)

The lower panel of Fig. 8.2 shows the relationship of information costs to σ computed for several bias levels as in the case of relevance coefficients. The higher the δ value, the lower is the level of P that can be obtained. At the maximum, the P curve shifts direction to a higher σ for highly biased results. The value of information will reach a maximum for σ between 5% and 7%, since over time further σ decreases prove to be too expensive.

In the case of INAA, if we have to determine several elements with high information content (high relevance), we should choose counting time that can achieve a satisfactory σ level for elements having the highest detection limit in the group. Otherwise, we obtain zero values for some of the information gain, and consequently a rather low value of information profitability [Eq. (8.2)]. In general, for INAA, an optimization of irradiation, decay, and counting times should be carried out.

When we need highly accurate results, that are both precise and bias-free, we turn to a special model for relevance coefficient calculation to obtain maximum information profitability. As Fig. 8.2 shows, highly precise results are expensive. It is not cost effective, in general, to measure with extremely high precision if a relatively high nonzero bias can occur.

8.2 OPTIMIZATION OF ANALYTICAL PROCEDURE

Once a suitable method is chosen, we can direct our attention to securing optimum working procedures and conditions in which the analyses will be performed. Accordingly we begin by establishing what factors have an effect on the results. That can be done based on our knowledge of the process underlying the analytical method, our experience with it, and so on (for the significance of the factors in a Plackett-Burman experiment, see Doerffel and Eckschlager 1981, ch. 6).

The *simplex method* is often used to obtain the optimum conditions; this method has been described in sufficient detail by Doerffel and Eckschlager (1981) and Eckschlager and Stepánek (1979). The procedure is carried out sequentially in that the next experiment is performed according to the result of the preceding experiment. Algorithms can be written so that the whole procedure is implemented by a computer.

The success of analytical optimization procedures depends on which objective function is chosen. Quantities that are too simple such as signal intensity may not be the most convenient; signal-to-noise ratio, information gain, or information profitability are better suited. These latter quantities are often referred to as *compromise target quantities*, which means that inherent in them are several single target properties such as precision, accuracy, and multicomponent suitability.

Recent advances in chemometrics have enabled direct multicriteria optimization (multivariate optimization, polyoptimization, target vector optimization) whereby m several target quantities can be optimized simultaneously by their dependence on p influence parameters: $(y_1, y_2, \ldots, y_m) = f(x_1, x_2, \ldots, x_p)$. For this purpose "classical" multivariate techniques like PLS (partial least squares) modeling (Wold et al. 1986; Wienke et al. 1989) and ORM (overlapping resolution maps; Laub and Purnell 1975; Sachok, Kong, and Deming 1980; Wienke and Danzer 1992) are used as well as modern strategies like *neural networks* (Long, Gregoriou, and Gemperline 1990; Zupan and Gasteiger 1991; Harrington 1993), and *genetic algorithms* (Lucasius and Kateman 1991; Li, Lucasius, and Kateman 1992).

8.3   INCREASE OF USEFUL INFORMATION AMOUNT IN INSTRUMENTAL ANALYSIS BY SIGNAL PROCESSING

Signal functions are obtained by instrumental methods of analysis as two-dimensional information, $y = f(z)$. The potential information amount of a signal is characterized by Eqs. (7.18) through (7.24), theoretically by the signal's resolution power N_p and its intensity resolution m_y which connects with the signal-to-noise ratio S/N according to Eq. (6.11). In instrumental-analytical practice we have

$$M = N_a \ \mathrm{lb} \frac{S}{N} \tag{8.4}$$

where N_a is the practical signal resolution. An increase of the information amount according to the Eq. (8.4) is possible in two ways: (1) by *improvement of signal resolution* and (2) by *increasing the signal-to-noise ratio* (see Danzer, Hopfe, and Marx 1982).

A signal function $y = f(z)$ is recorded over time by scanning the signal region z_{min} to z_{max}. From the standpoint of signal theory it is useful to regard a signal function as a time function, $y = f(t)$. This

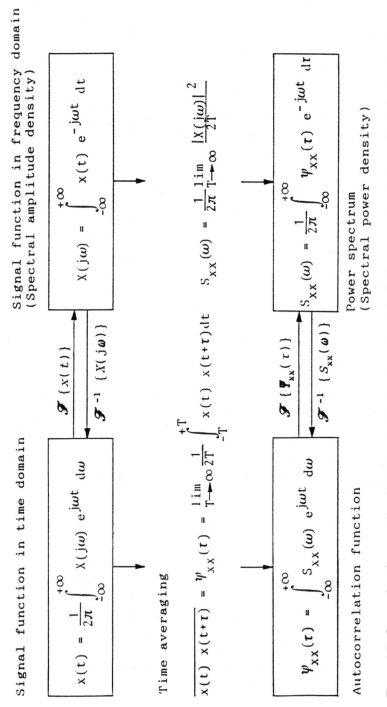

Figure 8.3. Connections between signal functions in the time domain and in the frequency domain (according to Woschni 1973. In analytical chemistry the power spectrum is mostly used as a function in the frequency domain; $\mathscr{F}$ represents a Fourier transform (Eqs. (8.5) to (8.10), from the top left-hand corner to the bottom right-hand corner)

174

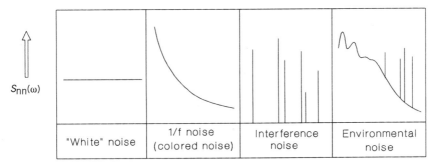

Figure 8.4. Schematic of the power of spectra common noise

function may be transformed from the *time domain* into the *frequency domain* (in general, from object domain into image domain) by means of transformations, such as the Fourier transform and the Laplace transform. Fourier transforms of signal functions in the time and frequency domains that are important in signal processing are shown in Fig. 8.3.

8.3.1   Increasing the Signal-to-Noise Ratio

Signal functions $f(t)$ can be regarded as consisting of the original *useful signal function*, $x(t)$, which is superposed by a noise function $n(t)$,

$$f(t) = x(t) + n(t) \qquad (8.11)$$

Noise behavior is generally additive; the types of noise are given in Fig. 8.4.

As a rule, in analytical instruments all types of noise can coexist —white and flicker noise $(1/f)$ as well as interferences. The following procedures can be applied to increase computationally the signal-to-noise ratio:

1. Signal averaging (accumulation)
2. Digital filtering
3. Autocorrelation
4. Smoothing of signal function

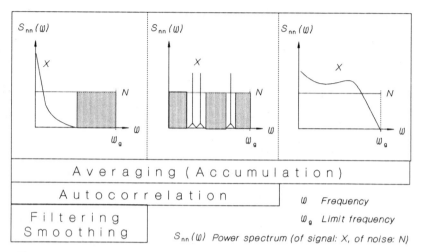

Figure 8.5. Characteristic relations between signal and noise power spectra

The applicability of these procedures depends on the type and form of the power spectra of signal and noise, respectively, particularly if there exist signal-free noise regions as shown in Fig. 8.5.

Signal averaging is the only method by which the S/N ratio can be increased universally. In the case where the signal function is repeatable and n measurements are averaged, the S/N ratio is increased by $\sqrt{n}$ according to

$$\left(\frac{S}{N}\right)_n = \frac{nx_i}{\sqrt{(ns_x^2)}} = \sqrt{n}\,\frac{x_i}{s_x} = \sqrt{n}\left(\frac{S}{N}\right)_i \qquad (8.12)$$

When the power spectrum of the signal function is limited to a certain frequency region (see Fig. 8.5, left and middle panels), filtering procedures can be applied to improve the S/N. The principle of filtering consists in a *convolution* of the noised signal function $f(t)$, see Eq. (8.11), with a filter function $g_F(t)$ whose transmission function $G_F(\omega)$ (i.e., the Fourier-transformed filter function) ideally becomes zero in the signal-free frequency regions.

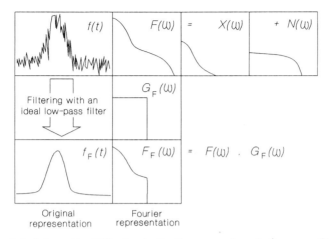

Figure 8.6. Schematic of filtering in the frequency domain (low-pass filtering)

For example, for the filtered signal function[1]

$$f_F(t) = f(t) * g_F(t) = x(t) * g_F(t) + n(t) * g_F(t) \quad (8.13)$$

the Fourier form is

$$F_F(\omega) = F(\omega) \cdot G_F(\omega) = X(\omega) \cdot G_F(\omega) + N(\omega) \cdot G_F(\omega) \quad (8.14)$$

The improvement of the signal-to-noise ratio is schematically shown in Fig. 8.6. As the figure indicates, an increase of S/N occurs if the transmission function of the filter $G_F(\omega)$ is limited.

Some other important procedures that have been effected by today's instruments are *optimum filtering*—where the frequency spectrum of the signal form function $X_s(\omega)$ itself is used as a filter function—and the Kalman filtering, which is a special recursive technique (see Rutan and Brown 1984; Brown 1986).

---

[1]The symbol $*$ means here the convolution operation which is carried out by integration (τ is an integration variable):

$$f(t) * g_F(t) = \int_{-\infty}^{+\infty} f(t - \tau) g_F(\tau) \, d\tau = \int_{-\infty}^{+\infty} f(\tau) g_F(t - \tau) \, d\tau$$

The *autocorrelation* technique is close in concept to optimum filtering. Since the true signal form function is seldom known, the signal function itself is used as a filter function, $G_F(\omega) = F(\omega)$. The *smoothing* of a signal function is carried out by polynomial least squares fitting. Some averaging procedures are similar to smoothing, particularly *boxcar averaging* and *moving window averaging*. The S/N improvement by these techniques is theoretically also $\sqrt{n}$, where n is the number of boxcar points (which in practice can be no more than five). More detailed discussion of signal-to-noise enhancement techniques can be found in Coors (1968), Hieftje (1972), Horlick (1972), Helstrom (1975), Dulaney (1975), and Sharaf, Illman, and Kowalski (1986).

8.3.2   Improvement of Signal Resolution

The true signal function is strongly influenced by the measuring system, namely in such a way that the signal resolution becomes worse. The true signal function cannot be observed or recorded in any way. Rather, the signal function is measured by a change in objective technical insufficiencies (e.g., the finite slit width in spectrometers). The change in a signal function by way of the system function is schematically shown in Fig. 8.7.

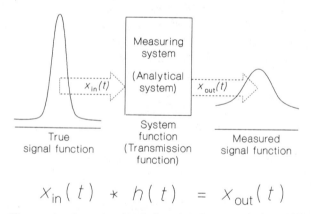

$$X_{in}(t) \; * \; h(t) \; = \; X_{out}(t)$$

Figure 8.7. Change in the true signal function by convolution with the system function during the measuring stage

Each measured signal function results from a convolution of the true signal function $x_{in}(t)$ with the transmission function of the analytical system (system function) $h(t)$. The analytical system shown in the figure comprises all the physical and technical subsystems of analytical instruments including excitation, electronics, and recording.

In the Fourier (frequency) domain the convolution is represented by multiplication:

$$X_{in}(\omega) \cdot H(\omega) = X_{out}(\omega) \tag{8.15}$$

Therefore the direct way to cancel convolution is by deconvolution. The following procedures show the different ways in which signal resolution can be improved:

1. *Deconvolution* of the measured signal function that refers to the system function, $X_{in}(\omega) = X_{out}(\omega)/H(\omega)$, yielding $X_{true}(\omega) = X_{meas}(\omega)/H(\omega)$. Although, in practice, the determination of the total system function is relatively difficult and expensive, in some cases deconvolution can be successfully applied to an essential partial system function (e.g., the slit function of spectrometers) (Kauppinen et al. 1981).

2. *Convolution* of the measured function with another suitable function (e.g., higher even derivations of the signal form function). Similar effects can be obtained by cross-correlation with Gaussian, Lorentzian, or rectangular signals (Horlick and Codding 1973).

3. *Differentiation* of signals. This procedure dates back to Gauss who found that the position of a higher deviation of a signal function can be determined more precisely than that of the original function because the zero passages of the odd deviations become steeper and the widths of the even deviations narrower (Morrey 1968; Maddams 1980). Differentiation is useful for the recognition of the components in complex signal functions.

4. *Curve fitting* by least squares minimization. Curve fitting is one of the most common techniques used to resolve overlapping signals. It is widely used in UV–VIS–, IR–, Raman– and

NMR spectroscopy as well as in chromatography where the peak shapes are modeled by linear superposition of Gaussian, Lorentzian, or Voigtian profiles. It is important to have some a priori information on the number of signals needed to reconstruct the original signal function (Kelly and Harris 1971; Vandeginste and DeGalan 1975; Maddams 1980).

5. Multivariate techniques related with *factor analysis*. This procedure is useful in extracting information on independent components in spectra and their mole fractions (e.g., by GC–MS) (Sharaf and Kowalski 1981, 1982); *rank annihilation* based on factor analysis can be applied successfully when the superposed signal functions are known a priori (Ho, Christian, Davidson 1978, 1981; Sanchez and Kowalski 1986).

Methods for increasing the signal-to-noise ratio, on the one hand, and for improving signal resolution, on the other hand, are based partly on inverse operations (convolution/deconvolution, integration/differentiation). Consequently S/N improving procedures may even worsen signal resolution, and vice versa. To ensure an improvement in both S/N and the resolution power, two or more procedures are combined; for example, averaging, smoothing, and curve fitting are frequently used together.

Extraction of useful information from a large amount of measured data (e.g., chemical information from analytical data) will be covered in Section 10.3.

8.4   SYSTEM BASIS FOR SEEKING THE OPTIMUM ANALYTICAL STRATEGY

The preceding sections gave a brief account of present analytical strategies for seeking an optimum that have, unfortunately, become too routine. In large part this is due to the fact that procedures are controlled by commercial computer software.

A somewhat different approach to understanding the problem of attaining a feasible optimum strategy is the system approach. While in the present approach the analytical method and the signal processing procedure are primarily optimized while the remaining partial operations are only taken into account to a greater or lesser extent, the system understanding of the analytical process requires its optimization in its entirety.

The optimization of a system, however, cannot be accomplished in a single operation. The first step is to select the optimum method; the second, to optimize the signal processing procedure; and the third, to transform the signal into analytical information. Subsequently the whole procedure—from sampling to weighting to sample decomposition or separation to the calculation of the result —is tested and, if appropriate, additionally optimized.

During this procedure it is useful to compare the expected information gain $I(r; p, p_0)_0$, determined a priori, and the actual information gain $\hat{I}(r; p, p_0)$, calculated from actual results. This relation is written

$$I(r; p, p_0)_0 - \tfrac{1}{2}t(\alpha, f)^2 \le \hat{I}(r; p, p_0) \le I(r; p, p_0)_0 \quad (8.16)$$

Equation (8.16) can be applied without much difficulty to the analysis of major contents. In trace analysis, however, we frequently can achieve no more than

$$0 \le \hat{I}(r; p, p_0) \le I(r; p, p_0)_0 \quad (8.17)$$

A property relevant in the assessment of an analytical procedure is its ruggedness; this term refers to the low "sensitivity" of the results to deviations from the optimum conditions in which the analysis is carried out. So far, ruggedness has not been defined quantitatively, but a simple criterion can be adopted: If the information gain $\hat{I}(r; p, p_0)$ changes very little and stays within narrow limits with minor deviations from the optimum procedure, the method can be considered rugged.

The final assessment of whether an optimum strategy has really been attained consists in an evaluation of how the procedure has stood the test in an interlaboratory experiment.

REFERENCES

Brown, S. D. 1986. *Anal. Chim. Acta* **181**, 1.
Coors, T. 1968. *J. Chem. Educat.* **45**, A533, A583.

Danzer, K., Eckschlager, K., and Wienke, D. 1987. *Fresenius Z. Anal. Chem.* **327**, 312.

Danzer, K., Hopfe, V., and Marx, G. 1982. *Z. Chem.* **22**, 332.

Decker, Jr., J. A. 1972. *Anal. Chem.* **44** (2), 127A.

Doerffel, K., and Eckschlager, K. 1981. *Optimale Strategien in der Analytik*. Deutscher Verlag für Grundstoffindustrie, Leipzig.

Doerffel, K., Eckschlager, K., and Henrion, G. 1990. *Chemometrische Strategien in der Analytik*. Deutscher Verlag für Grundstoffindustrie, Leipzig.

Dulaney, G. 1975. *Anal. Chem.* **47** (1), 24A.

Eckschlager, K., and Stepánek, V. 1979. *Information Theory as Applied to Chemical Analysis*. Wiley, New York.

Eckschlager, K., and Stepánek, V. 1985. *Analytical Measurement and Information*. Research Studies Press, Letchworth.

Esteban, M., Ruisanchez, I., Larrechi, M. S., and Ruis, F. X. 1992. *Anal. Chim. Acta* **268**, 95, 107.

Gerritsen, M., van Leeuwen, J. A., Vandeginste, B. G. M., Buydens, L., and Kateman, G. 1992. *Chemometrics Intell. Lab. Syst.* **15**, 171.

Hálová, J. 1989. *Chem. Prumysl* **39**, 659 (in Czech).

Harrington, P. de B. 1993. *Chemometrics Intell. Lab. Syst.* **18**, 157.

Helstrom, C. W. 1975. *Statistical Theory of Signal Detection*. Pergamon Press, Oxford.

Hieftje, G. M. 1972. *Anal. Chem.* **44** (6), 81A; (7), 69A.

Horlick, G. 1972. *Anal. Chem.* **44**, 943.

Horlick, G., and Codding, E. G. 1973. *Anal. Chem.* **45**, 1749.

Klaessens, J., and Kateman, G. 1987. *Fresenius Z. Anal. Chem.* **326**, 203.

Klaessens, J., van Beysterveldt, L., Saris, T., Vandeginste, B., and Kateman, G. 1989a. *Anal. Chim. Acta* **222**, 1.

Klaessens, J., Sanders, J., Vandeginste, B., and Kateman, G. 1989b. *Anal. Chim. Acta* **222**, 19.

Lahiri, S., and Stillman, M. J. 1992. *Anal. Chem.* **64**, 283A.

Laub, R. J., and Purnell, J. H. 1975. *J. Chromatogr.* **112**, 71.

Li, T. -H., Lucasius, C. B., and Kateman, G. 1992. *Anal. Chim. Acta* **268**, 123.

Long, R. J., Gregoriou, V. G., and Gemperline, P. J. 1990. *Anal. Chem.* **62**, 1791.

Lucasius, C. B., and Kateman, G. 1991. *Trends Anal. Chem.* **10**, 254.

Maddams, W. F. 1980. *Appl. Spectrosc.* **34**, 245.

Marshall, A., and Comisarow, M. B. 1975. *Anal. Chem.* **47**, 491A.

Morrey, J. R. 1968. *Anal. Chem.* **40**, 905.

Mulholland, M., Walker, N., van Leeuwen, J. A., Buydens, L., Maris, F., Hindriks, H., Schoenmakers, P. J., and Kateman, G. 1991. *Mikrochim. Acta* **II**, 493.

Obrusnik, I., and Eckschlager, K. 1993. *J. Radioanal. Nucl. Chem.* **169**, 347.

Pytela, O., Hálová, J., and Ludwig, M. 1990. *Collect. Czech. Chem. Commun.* **52**, 2629.

Roy, B. 1987. *Cahiers du Centre d'Études de Recherche Operationelle* **20**. Brussels.

Rutan, S. C., and Brown, S. D. 1984. *Anal. Chim. Acta* **160**, 99.

Sachok, B., Kong, R. C., and Deming, S. N. 1980. *J. Chromatogr.* **199**, 317.

Wienke, D., and Danzer, K. 1992. *Fresenius J. Anal. Chem.* **342**, 1.

Wienke, D., Danzer, K., Gitter, M., Aures, J., Münch, U., Byhan, H. G., and Pohl, H. J. 1989. *Anal. Chim. Acta* **223**, 247.

Wold, S., Sjöstrom, M., Carlson, R., Lundstedt, T., Hellberg, S., Skargerberg, B., Wikstrom, C., and Öhmann, J. 1986. *Anal. Chim. Acta* **191**, 17.

Woschni, E. -G. 1973. *Informationstechnik—Signal, System, Information.* Technik, Berlin.

Wünsch, G., and Gansen, M. 1989. *Fresenius Z. Anal. Chem.* **333**, 607.

Wünsch, G., and Schumnig, K. 1991. *Fresenius J. Anal. Chem.* **339**, 800.

Ziessow, D. 1973. *On-line Rechner in der Chemie.* de Gruyter, Berlin.

Zupan, J., and Gasteiger, J. 1991. *Anal. Chim. Acta* **248**, 1.

CHAPTER

9

QUALITY ASSURANCE

An increasing number of analyses in science, technology, and society involve not only the development of new and improved analytical procedures and techniques but also their supervision and control. In the 1970s the Kennedy Hearings on irregularities in chemical and toxicological investigations and reports led to the enactment of decrees for *Good Laboratory Practice* (*GLP*). These decrees were enacted to advance the quality of analytical investigations in general as well as in special fields (Department of Health, Education, and Welfare, Food and Drug Administration, 1978; Environmental Protection Agency 1979, 1980; OECD, 1981).

Today the principles of GLP represent a legal frame of quality assurance worldwide. The relation between GLP, on the one hand, and external (EQA) as well as internal quality assurance (IQA) is shown in Fig. 9.1. The main constituents of internal quality assurance are *Standard Operating Procedures* (*SOPs*), *Statistical Quality Control* (*SQC*), *Certified Reference Materials* (*CRMs*), and *Good Analytical Practice* (*GAP*). *Interlaboratory Studies* represent the most important tool of external quality assurance.

9.1 QUALITY CONTROL BY ANALYTICAL CHEMISTRY

Quality control of materials, products, goods, or ambient conditions is frequently exercised by analyzing one essential component or a few selected components. In analytical quality control we have to examine whether or not the content of the determined component lies within given tolerance limits.

In general, the relation between the result of analysis and the tolerance interval can be judged by the divergence measure of information, see Sections 3.8. and 6.6. To express the information content of the analytical quality control in this way, it is important

185

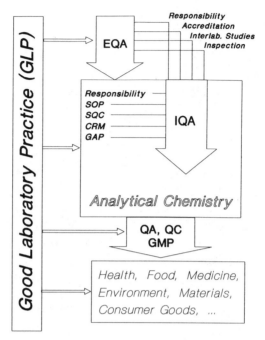

Figure 9.1. Components of external and internal quality assurance

to include in the a priori distribution the tolerance limits. We will discuss three possible ways of establishing tolerance limits; two of these have been characterized by Eckschlager (1978) as uniform distributions within the tolerance interval $\langle x_1, x_2 \rangle$ as shown in Fig. 9.2a and 9.2b. In general, the results can be found in the range $\langle x_0, x_3 \rangle$. In the normal case they lie within the standard range (tolerance interval) $\langle x_1, x_2 \rangle$. When tolerance limits are exceeded, there is a loss or reduction of quality. Therefore analyses are carried out to maintain the tolerance limits.

As we showed in Section 3.8, the information content of quality control can be estimated by the divergence measure (see Eckschlager 1978; Eckschlager and Stepánek 1979) as

$$I(p, p_0) = \int_{x_0}^{x_1} p(x) \mathrm{lb} \frac{p(x)}{p_2(x)} \, dx + \int_{x_1}^{x_2} p(x) \mathrm{lb} \frac{p(x)}{p_1(x)} \, dx$$

$$+ \int_{x_2}^{x_3} p(x) \mathrm{lb} \frac{p(x)}{p_3(x)} \, dx \tag{9.1}$$

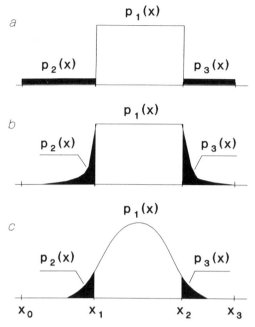

Figure 9.2. Different models of the distributions inside and outside of the tolerance limits

where the probability density $p_i(x)$ characterizes the a priori distribution and fulfills the equations

$$\int_{x_{i-1}}^{x_i} p_i(x)\, dx = \alpha_i$$

$$\sum_{i=1}^{3} \alpha_i = 1$$

According to Fig. 9.2, $p_1(x)$ is the distribution within, and $p_2(x)$ and $p_3(x)$ the distribution outside of the tolerance limits. The probability density function $p(x)$ characterizes the a posteriori distribution of the analytical result. The limits x_0 and x_3 must be such that

$$\int_{x_0}^{x_3} p_i(x)\, dx \approx 1.$$

In general, we consider a normal distribution of the results of analyses, where the mean is usually $\mu \geq x_0 + 3\sigma$ and $\mu \leq x_3 - 3\sigma$, so $p(x)$ for $x \leq x_0$ or $x \geq x_3$ is negligibly small. In analytical quality control the computed content of the analyte must lie within the tolerance limits x_1 and x_2, which are usually given, if the product under investigation is to be classified as satisfactory.

In defining the a priori distribution, we state that the content μ of the analyte lies in the interval $\langle x_1, x_2 \rangle$ with a probability α. In the panels (a) and (b) of Figure 9.2 the distribution is assumed to be uniform. With a probability $(1 - \alpha)$, the content μ lies outside of the tolerance interval: in the panel (a) with a constant probability, independent of its distance from the limits $x_1 - \mu$ or $\mu - x_2$; in panel (b) with a decreasing probability with increasing difference $x_1 - \mu$ or $\mu - x_2$ (these two cases and their given conditions are described by Eckschlager 1978). We now turn to the different models of information content that can be obtained as illustrated in Fig. 9.2.

First, for cases (a) and (b), where the found result lies within the tolerance limits, we have

$$I(p, p_0)_w = 1b \frac{x_2 - x_1}{\alpha\sigma\sqrt{(2\pi e)}} \tag{9.2}$$

Then, for case (a) outside of the tolerance limits, we have

$$I(p, p_0)_{ou} = 1b \frac{(x_1 - x_0) + (x_3 - x_2)}{(1 - \alpha)\sigma\sqrt{(2\pi e)}} \tag{9.3}$$

and for case (b) outside of the tolerance limits,

$$I(p, p_0)_{ou} = 1b \frac{x_2 - x_1}{\alpha\sigma\sqrt{(2\pi e)}} + \frac{\alpha}{1 - \alpha} A \tag{9.4}$$

with $A = (x_1 - \mu)/(x_2 - x_1)$ for $\mu < x_1$ and $A = (\mu - x_2)/(x_2 - x_1)$ for $\mu > x_2$.

Note that model (a) gives constant information content for analytical results both within and beyond the tolerance limits because α is constant but that in case (b) the results outside of the

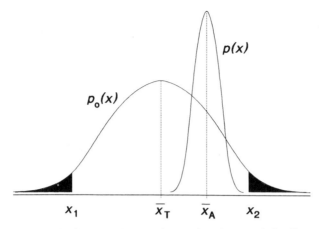

Figure 9.3. Normal distributed tolerance interval and normal distributed results of analytical quality control

tolerance range are increasing in information content the more the distance increases between μ and the limit x_1 or x_2 grows (the numerator A increases).

Both distribution models (a) and (b) have the disadvantage that scattering results *within* the tolerance range are characterized by the same information content; it is well-known in quality control (e.g., from control charts) that an approach toward tolerance limits constitutes a warning. Therefore it is no surprise that the information content is low if the result of the analytical quality control is close to the target value $\bar{x}_T$ and high if the result comes close to the tolerance limits. In practice, in analytical quality control there is only a small difference in information when the result is near the limit or slight beyond the limit. These existing conditions can be considered best if we apply a Gaussian a priori distribution $p_0(x)$ with the tolerance limits x_1 and x_2, as shown in Fig. 9.3.

We will give some examples that illustrate possible situations in quality control where normal distributed tolerance limits are assumed; the standard deviation would be one-fifth of that of process, $\sigma = \sigma_0/5$.

Figure 9.4a shows a random scattering of results about the target value. The results are normalized so that the target value $\bar{x}_T = 0$, the upper warning limit $x_{uw1} = 2$, the upper tolerance (action) limit

$x_{ut1} = x_2 = 3$, the lower warning limit $x_{1w1} = -2$, and the lower tolerance limit $x_{1t1} = x_1 = -3$. In Fig. 9.4b the results slope where the warning limit has been reached. Represented as well are the so-called CUSUM (cumulative sum) values which are frequently used in process control and quality control of products. (Page 1954; Barnard 1959; Ewan and Kemp 1960). At last the information content of the results is given, both as individual and as sum, where the information content of the individual result is calculated accord-

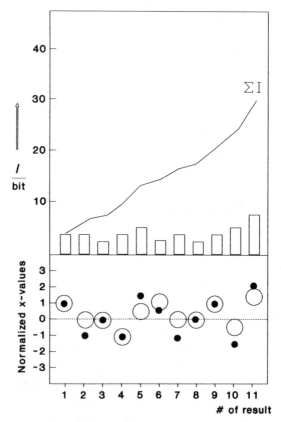

Figure 9.4. Different situations in quality control: (*left*) randomly scattering values, (*right*) trend-affected results, where ● = results of analyses, ○ = CUSUM, bars = information content of each result, and ΣI = information amount of the sequence of results

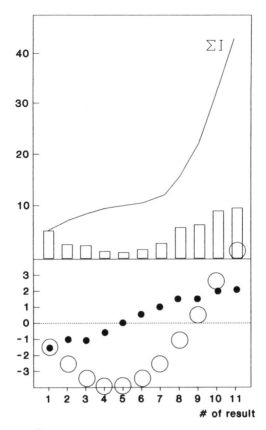

Figure 9.4. (*Continued*)

ing to Kullback (1959) (see also Eckschlager and Stepánek 1979):

$$I(p, p_0) = 1b\frac{\sigma_0}{\sigma} + \kappa\frac{(\bar{x}_T - \bar{x}_A)^2 + \sigma^2 - \sigma_0^2}{\sigma_0^2} \tag{9.5}$$

Both panels of Fig. 9.4 are represented in Fig. 9.2 as cases (*a*) and (*b*), and by Eq. (9.2) they yield a constant information content for all results 1 to 11.

However, for case (*c*) of Fig. 9.2, the information content $I(p, p_0)$ in Eq. (9.5) increases as the result of analysis moves away from the

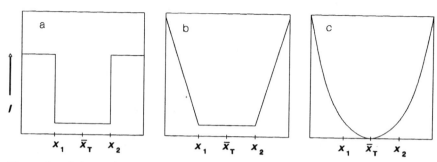

Figure 9.5. Information content for three different models of an a priori distribution; cases *a*, *b*, and *c* of Fig. 9.2

target value. The differences between the information-theoretic nature of cases (*a*), (*b*), and (*c*) are shown in Fig. 9.5.

From model (*c*) and Eq. (9.5), we see that as the warning limit $(2\sigma_0)$ is reached, the information content becomes

$$I(p, p_0)_{w1} = 1b\frac{\sigma_0}{\sigma} + \frac{\kappa(3\sigma_0^2 + \sigma^2)}{\sigma_0^2} \qquad (9.6a)$$

and the tolerance limit $(3\sigma_0)$ becomes

$$I(p, p_0)_{t1} = 1b\frac{\sigma_0}{\sigma} + \frac{\kappa(8\sigma_0^2 + \sigma^2)}{\sigma_0^2} \qquad (9.6b)$$

The advantage to using model (*c*) for characterizing the information of analytical quality control is that trends in the data are indicated by information content in the same way as CUSUM. Long increases and decreases in successive results show systematic changes in the control results. This is apparent in Fig. 9.4 in the comparison of random and trend results.

The width of the tolerance limit can be estimated by a *tolerance coefficient* k (Hald 1952). The tolerance coefficient depends on the probability α of the a priori distribution of the process and on γ values so that $100.\gamma\%$ of all products pass the tolerance limits with the probability α. The tolerance coefficients

$$k = \frac{x_2 - x_1}{2\sigma_0} \tag{9.7}$$

for a normally distributed quantity have been tabulated (Hald 1952, p. 315). The tolerance coefficients characterize the total variability of content $\bar{x}_A$ of a determined component (found by the given analytical method) on a larger number of charges, and it is naturally due not only to the inaccuracy of the analytical results but also to a scatter caused by the manufacturing process (γ). Therefore the standard deviation of the mean value $\bar{x}_A$ found from n_A parallel determinations $\sigma_{\bar{x}}$, which characterizes the accuracy of the analytical results, will be always smaller than σ_0. Then it always is $2k \leq (x_2 - x_1)/\sigma_{\bar{x}}$, and since $\sigma_{\bar{x}} = \sigma/\sqrt{n_A}$, we have $(x_2 - x_1)\sqrt{n_A}/\sigma \geq 2k$. This inequality not only limits the width of the tolerance limits with respect to the standard deviation of the a posteriori distribution but also determines the minimum information content of the analysis that can be used in the analytical quality control as

$$I(p, p_0)_{min} = 1b\frac{2k\sqrt{n_A}}{\alpha\sqrt{(2\pi e)}} \tag{9.8}$$

The information content of analytical results is independent of whether the quality control is carried out continually. Process control as continual quality control (Leemans 1971) can also be characterized by information theory (Zettler 1969; Danzer 1978), since principles of time-series analysis (Doerffel, Niedner, and Raschke 1990) auto- and cross-correlation analysis (see Section 8.3; Leemans 1971) can be applied.

9.2 QUALITY ASSURANCE WITHIN ANALYTICAL CHEMISTRY

Quality assurance is necessary in analytical chemistry, and it must be carried with the reliability. Quality assurance is characterized by both *internal* and *external* measures. These are four essential tasks to internal quality assurance:

1. Choice and validation of analytical methods (Taylor 1980).
2. Permanent control of key parameters of analytical methods, recalibration.
3. Permanent supervision of the staff by control analyses.
4. Application of reference materials.

We discussed earlier the most important of these tasks as they relate to the choice and optimization of analytical methods (Section 8.1) and the use of standard reference materials (Section 6.4). The principles of permanent control of analytical methods are the same as that of quality control in general (Section 9.1), and they are carried out frequently by use of control charts. However, external quality control almost exclusively involves *proficiency testing* by *interlaboratory comparisons*.

Interlaboratory comparisons are carried out for different purposes. According to Horwitz (1989), the five different aims of interlaboratory studies, each characterized by its own statistical assumptions and models, are the following:

1. *Collaborative studies* to determine the performance characteristics of a method of analysis.
2. *Comparative studies* to compare the performance of several methods of analysis.
3. *Proficiency studies* to determine the performance of analysts or laboratories.
4. *Consensus studies* to determine a value for use as an estimate of the "true values."
5. *Certification studies* to assign the "true value" of a material with a stated uncertainty.

Whereas aims 1 and 2 are directed to methods and aims 4 and 5 focus on the material, aim 3 on proficiency studies concerns external quality assurance, particularly accreditation of laboratories such as the AOCS Certified Laboratory Program. There are several papers on design, conduct, evaluation, and interpretation of interlaboratory studies (ISO 1983–1986, 1989; ISO/REMCO 1991; IUPAC 1988).

The evaluation and assessment of laboratories is carried out in analytical practice by different quantities: by the absolute deviation of the laboratory mean $\bar{x}_1$ from the "true value" or target value μ, by $|\bar{x}_1 - \mu|$, and by relative measures such as z-scores that represent normalized laboratory deviations, $z_1 = (\bar{x}_1 - \mu)/\sigma$ (Thompson 1992), or similar quantities (Gorski and Bobrowski 1989).

Information theory is used in expert systems for the evaluation and interpretation of interlaboratory comparisons (Danzer, Wank, and Wienke 1991) in order to evaluate individual results (or methods) of single laboratories in relation to a target value. The basic measure of inaccuracy is that of Kerridge and Bongard derived in Section 6.2. (see Danzer, Eckschlager, and Wienke 1987):

$$I(r; p, p_0) = 1b\frac{x_2 - x_1}{\sigma\sqrt{(2\pi)}} - \kappa\frac{\delta^2 - \sigma^2}{\sigma^2} \qquad (9.9)$$

In applying model (9.9) for interlaboratory studies that determine the performance of methods, laboratories, or analysts, we must always know what is our "true" target value μ_t. In the case of proficiency testing, we have for the information contribution of the ith laboratory (or method)

$$\hat{I}(r; p, p_0)_i = 1b\frac{x_2 - x_1}{2s_i t_{a, f}/\sqrt{n_i}} - \kappa\frac{(\bar{x}_i - \mu_t)^2 - s_i^2}{s_i^2} \qquad (9.10)$$

where x_2 is the upper limit, x_1 the lower limit of the expectation range of the results, and $\bar{x}_i$, s_i, and n_i are the mean,, standard deviation, and number of replications, respectively, of the ith laboratory; $t_{a, \infty} = \frac{1}{2}\sqrt{(2\pi)}$. If no other reasons exist, it may be useful to set $x_2 = \mu_t + \mu_t/2$ and $x_1 = \mu_t - \mu_t/2$, and therefore $(x_2 - x_1) = \mu_t$ for the main components. For trace analyses it is not certain that $\mu_t > 2s_i t_{a, f}\sqrt{n_i}$ from the beginning. Besides, since $x_1 \ll x_2$, we can set $x_1 = 0$. So x_2 should be chosen in a way that $x_2 \approx 100\mu_t$.

The first term in Eq. (9.10) characterizes the precision of the laboratory with regard to a given working range $x_1 \ldots x_2$. However, the second term is determined by the relative inaccuracy of the actual result in relation to the target value μ_t. Large systematic errors cause a significant decrease of the information contribution, especially in the case of highly precise results.

From the theoretical as well as from the practical point of view (Eckschlager and Stepánek 1985; Danzer, Eckschlager, and Wienke 1987) the information contribution must become

$$\hat{I}_i = \begin{cases} \hat{I}(r;p,p_0)_i & \text{if } \hat{I}(r;p,p_0)_i \geq 0 \\ 0 & \text{if } \hat{I}(r;p,p_0)_i < 0 \end{cases} \qquad (9.11)$$

When a true value is not known but has to be determined by an interlaboratory study (consensus and certification study) in order to establish it for a reference material, then the inaccuracy term of Eq. (9.10) must be modified in the following way:

$$\hat{I}(r;p,p_0)_i = 1b \frac{x_2 - x_1}{2 s_i t_{a,f}/\sqrt{n_i}} - \kappa \frac{(\bar{x}_i - \bar{\bar{x}}_t)^2 + s_i^2 - s_i^2}{s_t^2} \qquad (9.12)$$

where $\bar{\bar{x}}_t$ is the total mean of the interlaboratory study (without outlying laboratory means, which have to be tested carefully) and s_t the standard deviation.

For multivariate problems—such as those concerning interlaboratory multicomponent comparisons and simultaneous consideration of several methods or laboratories—a multivariate measure of information has been developed (Danzer 1991; Dörfer 1990):

$$\begin{aligned} I(r;p,p_0)^m &= \int_{\mathbf{R}^m} r(\underline{x}) 1b \frac{p(\underline{x})}{p_0(\underline{x})} \, d\underline{x} \\ &= 1b \frac{\prod_{i=1}^{m}(x_{i2} - x_{i1})}{\sqrt{[(2\pi e)^m \det \Sigma]}} \\ &\quad - \kappa\left[(\underline{\mu} - \underline{x}_t)^T \Sigma^{-1}(\underline{\mu} - \underline{x}_t) + m\right] \\ &\quad + 1b\left[\chi_{[x_1, x_2]^m}(\underline{x})\right] \end{aligned} \qquad (9.13)$$

where m is the number of components and n the number of groups (laboratories, methods); with $i = 1 \ldots n$ we have

$\underline{x}^T = (\bar{x}_{1i}, \bar{x}_{2i}, \ldots, \bar{x}_{mi})$ the vector of group means
$\underline{x}_t^T = (\bar{\bar{x}}_1, \bar{\bar{x}}_2, \ldots, \bar{\bar{x}}_m)$ the vector of total means
$\underline{\mu}^T = (\mu_1, \mu_2, \ldots, \mu_2)$ the vector of true values
$\underline{\underline{\Sigma}}$ = variance/covariance matrix

The multivariate distributions are the following:

$$r(\underline{x}) = \left[(2\pi)^m \det \underline{\underline{\Sigma}}_\mu\right]^{-1/2} \exp\left[-\tfrac{1}{2}(\underline{x} - \underline{\mu})^T \underline{\underline{\Sigma}}_\mu^{-1}(\underline{x} - \underline{\mu})\right]$$

$$p_0(\underline{x}) = \frac{1}{\prod_{i=1}^m (x_{i2} - x_{i1})} \mathbf{\chi}_{[x_1, x_2]^m}(\underline{x})$$

$$p(\underline{x}) = \left[(2\pi)^m \det \underline{\underline{\Sigma}}\right]^{-1/2} \exp\left[-\tfrac{1}{2}(\underline{x} - \underline{\mu}_t)^T \underline{\underline{\Sigma}}^{-1}(\underline{x} - \underline{\mu}_t)\right]$$

$$\mathbf{\chi}_{[x_1, x_2]^m}(\underline{x}) = \begin{cases} 1 & \text{if } x_i \in [x_1, x_2] \\ 0 & \text{otherwise} \end{cases}$$

From the multivariate information gain $I(r; p, p_0)^m$ defined in a space $\mathbf{R}^m$, we can evaluate and interpret problems that are determined by correlated and interacting features (Danzer 1991; Wienke, Dörfer, and Danzer 1993).

9.3 INTERLABORATORY COMPARISONS

Our first example deals with an interlaboratory study for comparison of several methods. A round-robin test was carried out to analyze an aqueous nickel solution (Ohls and Sommer 1982) with a (true) target concentration $\mu = 500.0$ μg/ml Ni. A sample with this concentration was distributed to six laboratories for analysis by different methods. In each laboratory ten repeated determinations were carried out.

The means and standard deviations for each method are given in Table 9.1. Note that the information in the right-hand column not only results from the bias itself but also from its relation to the error of the method. A false result is more weighty, the more

Table 9.1. Information Contents Computed Univariately by Eq. (9.10)
with $x_2 - x_1 = 35$ μg / ml and $n_i = 10$

Method	$\bar{x}/(\mu g/ml)$	$s/(\mu g/ml)$	$\delta/(\mu g/ml)$	I/bit
Gravimetry	494.7	2.76	5.3	1.19
Titration	495.2	1.25	4.8	0 [a]
Polarography	501.4	4.80	1.4	2.94
Photometry	499.5	3.20	0.5	3.62
AAS	493.2	2.96	6.8	0 [b]
ICP–OES	502.2	5.46	2.2	2.75

[a] Formally calculated value: -5.66 bit.

[b] Formally calculated value: -0.06 bit.

precise the value is. The outcome of this example is plausible: The most accurate methods obtain the highest information contribution (photometry, polarography, and ICP-OES). However, the most precise methods give low or zero information, respectively, because they have low tolerance for inaccuracy; the confidence interval of the result does not include the true value.

Now let us consider this example from another standpoint. Let us assume that an interlaboratory study will be carried out to determine the estimate of the "true value." Then we have to decide which of the results should be included in the evaluation. If we proceed in the classical univariate way, we would sum the individual information contents: $M = \Sigma I_i$, in contrast to the multivariate point of view where we would apply a multivariate measure of information [Eq. (9.13)]. The results are compared in Table 9.2.

As can be seen from Table 9.2, the outcomes of the univariate and multivariate measures of information differ. The multivariate measure is zero for all combinations that include titration. The high misinformation due to titration affects the multivariate measure more than the univariate measure. But we must consider that Eq. (9.11) has been used for the calculation of the univariate information amount rather than Eq. (9.10), which would have a negative effect. In general, the multivariate measure is better suited than the univariate for problems where objects or features are considered jointly and interactions or correlations cannot be excluded.

Our second example refers to a proficiency study for examining the reliability of nine laboratories in the determination of boric acid

**Table 9.2. Comparison of the Univariate Information Amount and
the Multivariate Measure of Information (in Bits) in
Case of Combination and Unification of Different Results of Analysis**

Combination of Methods	$M_{uv} = \Sigma I_i$	$I(r; p, p_0)^m$
Gravimetry + titration	1.2	0
Titration + polarography	2.9	0
Polarography + photometry	6.6	8.6
Photometry + ICP	6.4	8.4
Photometry + AAS + ICP	6.4	9.5
Polarography + photometry + AAS	6.6	9.0
Polarography + photometry + ICP	9.3	12.6
Gravimetry + titration + polarography + photometry	7.8	0
Gravimetry + polarography + photometry + ICP	*10.5*	*14.6*
Gravimetry + titration + polarography + photometry + AAS + ICP	10.5	0

in a preservative used in treated timber. Six samples were distributed with the graduated (true) target values: 0.040, 0.080, 0.120, 0.160, 0.200, and 0.300 wt-%. Each laboratory carried out three repetitive measurements.

The results of the proficiency study are given in Table 9.3 together with the individual information content I_i of each laboratory. The latter depends on the accuracy and precision as expected. If we consider the information amounts $M = \Sigma^6 I_i$ of the laboratories, we find the following ranking list:

A: 46.62 bit

G: 40.46 bit

C: 38.46 bit

E: 37.52 bit

H: 35.01 bit

I: 33.65 bit

D: 30.01 bit

B: 26.06 bit

F: 15.54 bit

Table 9.3. Information Content (in Bits) of a Proficiency Study of Boric Acid in Preservative Treated Timber

Laboratory	Mean	Standard Deviation	I	Mean	Standard Deviation	I	Mean	Standard Deviation	I
A	0.0493	0.00379	6.32	0.085	0.001	8.63	0.115	0.0131	4.89
B	0.038	0.01646	4.59	0.0796	0.00929	4.76	0.112	0.0115	4.99
C	0.0396	0.00416	6.59	0.077	0.00265	5.89	0.116	0.00265	7.21
D	0.053	0.00173	7.06	0.0773	0.00666	4.65	0.124	0.0140	4.81
E	0.486	0.00666	5.56	0.0833	0.00208	7.47	0.121	0.00346	6.86
F	0.0295	0.04172	0.89	0.074	0.01217	2.63	0.096	0.0530	1.71
G	0.0476	0.00462	6.17	0.0876	0.00289	7.07	0.1273	0.00208	7.48
H	0.053	0.00656	5.14	0.081	0.00519	5.85	0.1233	0.00473	6.39
I	0.049	0.00346	6.48	0.082	0.00346	6.59	0.120	0.0060	6.07

Laboratory	Mean	Standard Deviation	I	Mean	Standard Deviation	I	Mean	Standard Deviation	I
A	0.1636	0.00153	8.00	0.198	0.0001	11.96	0.288	0.00265	6.82
B	0.1513	0.00586	5.89	0.1806	0.00404	5.83	0.254	0.01629	0
C	0.1543	0.00306	6.95	0.1926	0.00306	6.92	0.288	0.010	4.90
D	0.149	0.0020	7.31	0.1796	0.01115	4.28	0.2716	0.01106	2.80
E	0.1643	0.00586	6.05	0.200	0.0070	5.84	0.2876	0.00551	5.74
F	0.1626	0.04245	3.23	0.2383	0.00896	2.33	0.295	0.01418	4.75
G	0.167	0.00361	6.66	0.2056	0.00493	6.28	0.301	0.00361	6.80
H	0.169	0.00781	5.46	0.1976	0.00702	5.82	0.292	0.00436	6.34
I	0.183	0.0060	4.56	0.215	0.01249	4.53	0.278	0.00346	5.42

Source: Data from Dawson et al. (1990).

Note: Means and standard deviations in wt-%; $x_2 - x_1 = 1.000$ wt-%.

200

Laboratory F produced the most deviating results and therefore made the smallest information contribution. The procedure of determination of boric acid should be checked by this laboratory.

Our final example deals with an interlaboratory study for quality assurance: the analysis of fat in milk by 38 laboratories, some of which applied different methods. The data were generously provided to Carl (1992). The results are given in Table 9.4. In cases where the values of single determinations are equal, the standard deviation was calculated from a uniform distribution within the rounding range $s^2 = (x_{max} - x_{min})^2/12$. For example, for laboratories 1 and 6.2, the results are

$$x_1 = x_2 = 16.0$$

$$s = \sqrt{\frac{(16.05 - 15.95)^2}{12}} = 0.0289 \quad \text{(lab. 1)}$$

$$x_1 = x_2 = 15.97$$

$$s = \sqrt{\frac{(15.975 - 15.965)^2}{12}} = 0.0029 \quad \text{(lab. 6.2)}$$

The interlaboratory comparison is shown in Fig. 9.6.

The maximum information content belongs to the most accurate and most precise methods with respect to the representative overall mean $\bar{\bar{x}}_{tot} = 15.96$ wt-% ($s_{tot} = 0.30$ wt-%). This total mean was calculated after rejection of three outlying laboratory means, namely those of laboratories 28, 38, and 21 according to the outlier test of Graf and Henning (1952). The rejection is fully in agreement with the representation of data in Fig. 9.6 and with the information-theoretic evaluation because the information contributions for these three laboratories are equal to zero (formally negative).

Information theory enables us to evaluate and validate analytical methods and results of chemical analyses. The interlaboratory studies in quality assurance and accreditation depended on the information measure of inaccuracy developed by Kerridge and Bongard. Thus calculated the information provides a useful measure of the proficiency of methods, laboratories, and staff.

Table 9.4. Results (in wt-%) and Information Contributions (in Bits) of an Interlaboratory Study for Fat Determination in Milk

Laboratory Code	Mean	Standard Deviation	n	I	
1	16.0	0.0289	2	5.65	
2	15.5	0.0289	2	3.97	
3	15.5	0.1826	4	3.81	
4.1	15.867	0.0577	3	6.45	
4.2	16.1	0.0289	2	5.51	
5	15.953	0.0814	4	6.67	
6.1	16.025	0.0071	2	7.66 ■	
6.2	15.97	0.0029	2	6.66	
7.1	16.15	0.0029	2	6.38	
7.2	15.8	0.1414	2	3.17	
8	15.55	0.300	4	3.44	
9	16.115	0.0495	2	4.69	
10	16.2	—	1	—	
12	16.19	0.0029	2	6.24	
13	16.02	0.0566	2	4.67	
14	16.25	0.0566	2	4.02	
15	16.195	0.0071	2	7.25 ■	
16	15.85	0.0141	2	6.60	
17	15.797	0.0451	3	6.66	
18	16.15	0.0707	2	4.08	
19.1	15.29	0.0283	2	2.10	
19.2	15.635	0.0919	2	3.15	
20	16.12	0.0566	2	4.49	
21	16.875	0.1768	2	0	[−3.66]
22	16.067	0.0577	3	6.43	
23.1	16.235	0.0354	2	4.77	
23.2	16.155	0.0212	2	5.81	
24	15.645	0.0551	4	6.44	
25	16.155	0.0354	2	5.07	
26	16.15	0.0707	2	4.08	
27	16.28	0.0282	2	4.88	
28	14.20	0.0289	2	0	[−19.16]
29	16.48	0.0141	2	4.53	
30	16.145	0.0495	2	4.61	
31	16.0	0.0289	2	5.65	
33	15.915	0.0071	2	7.67 ■	
34	15.90	0.0566	2	4.67	
35	16.005	0.0778	2	4.22	
36	16.0	0.0289	2	5.65	
37	16.225	0.0212	2	5.55	
38	14.82	0.1838	2	0	[−7.42]
39	16.265	0.0778	2	3.49	
40	16.01	0.0424	2	5.09	

Source: Danzer (1993).

Fat content in %

Figure 9.6. Interlaboratory study of fat determination in milk

REFERENCES

Barnard, G. A. 1959. *J. Roy. Stat. Soc. Ser. B* **21**, 239.

Carl, M. 1991. Unpublished data.

Danzer, K. 1978. *Z. Chem.* **18**, 104.

Danzer, K. 1991. Information Theory—May It Be Useful for the Analyst? 42. Pittsburgh Conference, Chicago, Book of Abstracts #569.

Danzer, K., Wank, U., and Wienke, D. 1991. *Chemometr. Intell. Lab. Syst.* **12**, 69.

Danzer, K., Eckschlager, K., and Wienke, D. 1987. *Fresenius Z. Anal. Chem.* **327**, 312.

Danzer, K. 1993. Information Content of an Analytical Result and Its Assessment by interlaboratory Study. Ch. 21 in: *Analytical Quality Assurance and Good Laboratory Practice in Diary Laboratories* (Proc. Internat. Seminar AOAC Intern., CEC, IDF, VDM, Sonthofen 1992), Brussels.

Department of Health, Education, and Welfare, Food and Drug Administration. 1978. *Nonclinical Laboratory Studies, Good Laboratory Practice, Regulations.* Federal Register **43** (247), 59986.

Dörfer, J. 1990. Diploma thesis. Friedrich Schiller University, Jena.

Doerffel, K., Niedtner, R., and Raschke, U. 1990. *Anal. Chim. Acta* **238**, 55.

Eckschlager, K. 1978. *Coll. Czech. Chem. Commun.* **43**, 231.

Eckschlager, K., and Stepánek, V. 1979. *Information Theory as Applied to Chemical Analysis.* Wiley, New York.

Eckschlager, K., and Stepánek, V. 1985. *Analytical Measurement and Information.* Research Studies Press, Letchworth.

Environmental Protection Agency. 1979. *Good Laboratory Practice, Standards for Health Effects.* Federal Register **44** (91), 27362.

Environmental Protection Agency. 1980. *Proposed Good Laboratory Practice, Standards for Physical, Chemical, Persistence and Ecological Effects Testing.* Federal Register **45** (227), 77353.

Ewan, W. D., and Kemp, K. W. 1969. *Biometrika* **47**, 363.

Gorski, L., and Bobrowski, M. 1989. Wiss. Beitr. Friedrich Schiller University, Jena. COMPANA '88, 4. Tagung Computereinsatz in der Analytik, 74.

Graf, U., and Henning, H. J. 1952. *Mitt.-bl. Math. Stat.* **4**, 1.

Hald, A. 1952. *Statistical Theory with Engineering Applications.* Wiley, New York.

Horwitz, W. 1989. *J. Assoc. Off. Anal. Chem.* **72**, 145.

ISO. 1983. *Guide 38. General Requirements for the Acceptance of Testing Laboratories.* ISO, Geneva.

ISO. 1984. *Guide 43. Development and Operation of Laboratory Proficiency Testing.* ISO, Geneva.

ISO. 1985. *Guide 45. Guidelines for the Presentation of Test Results.* ISO, Geneva.

ISO 5725. 1986. *Precision of Test Methods—Determination of Repeatability and Reproducibility for a Standard Test Method by Interlaboratory Tests.* 2d ed. ISO, Geneva.

ISO. 1989. *Guide 35. Certification of Reference Materials—General and Statistical Principles.* 2d ed. ISO, Geneva.

ISO/REMCO. 1991. Proc. Fourth International Symposium on the Harmonization of Quality Assurance Systems in Chemical Analysis. ISO, Geneva.

IUPAC, Analytical, Applied and Clinical Chemistry Divisions. Interdivisional Working Party for Harmonization of Quality Assurance Schemes for Analytical Laboratories. 1988. *Protocol for the Design, Conduct and Interpretation of Collaborative Studies* (prepared for publication by W. Horwitz). *Pure Appl. Chem.* **60**, 855.

Kullback, S. 1959. *Information Theory and Statistics*. Wiley, New York.

Leemans, F. A. 1971. *Anal. Chem.* **43**(11), 36A.

OECD. 1981. *Decision of the OECD Council Concerning the Mutual Acceptance of Data in the Assessment of Chemicals*. Annex 1: *OECD Test Guidelines*; Annex 2: *Principles of Good Laboratory Practice*. C (81) 30 (Final).

Ohls, K., and Sommer, D. 1982. *Fresenius Z. Anal. Chem.* **312**, 195.

Page, E. S. 1954. *Biometrika* **41**, 100.

Taylor, K. 1983. *Anal. Chem.* **55**, 600A.

Thompson, M. 1992. *Anal. Proc.* **29**, 190.

Wienke, D., Dörfer, J., and Danzer, K. 1993. In preparation.

Woodwards, R. H., and Goldsmith, P. L. 1964. *Cumulative Sum Techniques*. Oliver & Boyd, Edinburgh.

Zettler, H. 1969. *Fresenius Z. Anal. Chem.* **245**, 1.

CHAPTER

10

DISTRIBUTION ANALYSIS: MICROANALYSIS, SURFACE ANALYSIS, AND SCANNING METHODS

Distribution analyses yield multidimensional information that can be described by mathematical functions comprising at least one variable spatial coordinate of the sample serving as an independent variable (Danzer 1974; Danzer et al. 1988). Figure 10.1 gives a schematic of distribution-analytical procedures and problems and their related applications (see Danzer, Schubert, and Liebich 1991; Ehrlich, Danzer, and Liebich 1979).

Local analysis (*or point analysis*) has a key position in distribution analysis because the procedures plane and volume analyses are derived from it by a joining of local analyses (points) in one to three directions of space ($1.0 \rightarrow 1.1 \rightarrow 1.2 \rightarrow 1.3$). By extending the "subsample" in one or more additional directions, we have the method of incomplete or partial integral distribution analysis, which results from local analysis ($2.1 \rightarrow 2.2 \rightarrow 2.3, 2.1 \rightarrow 3.2 \rightarrow 3.3$). Distribution analyses in a narrow sense are carried out practically by successive point analyses, namely as line scans (concentration profiles; see Fig. 10.2a), as surface scans (element images; see Fig. 10.3; or two-dimensional concentration profiles; see Fig. 10.4), and as volume analyses (volume scans, 3D images; see Fig. 10.5).

The limiting cases in Fig. 10.1 are represented by *testing of homogeneity* (1.2^* and 1.3^*) and *average analysis* (4.3). The latter does not belong to distribution analysis and is included only for the sake of completeness. In the lowest register of Fig. 10.1 we show the change in the crystal structure, which can be understood as a distribution analysis in dimensions of atoms.

Before we can describe and evaluate distribution-analytical procedures, we must consider *geometrical resolving power A* (see Danzer 1974; Danzer et al. 1988). We will include in our discussion known characteristics of related procedures such as precision (*concentration resolving power*) and *analytical* (*signal*) *resolving power*.

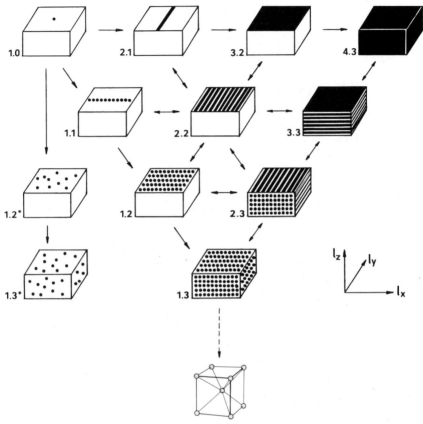

Figure 10.1. Systematic of distribution-analytical problems and procedures. The numbers correspond to equations in Table 10.1

10.1　GEOMETRICAL RESOLVING POWER

The potential geometric resolving power A^* refers to an ideal case where the maximum number of measuring points determined, e.g., by a fixed electron beam diameter with a given excitation voltage are independent of each other. We will describe this ideal geometric resolving power for three specific cases: lateral resolving power A_l^*, plane resolving power A_p^*, and volume resolving power A_v^*.

Lateral resolving power A_1^* is written

$$A_l^* = \frac{l_i}{\Delta l_i} \qquad (10.1)$$

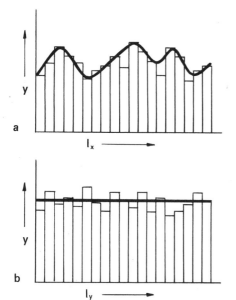

a

b

Figure 10.2. Examples for lateral distribution analysis: line scan (*a*), see 1.1, and integral lateral analysis (*b*), see 2.1, in Table 10.1

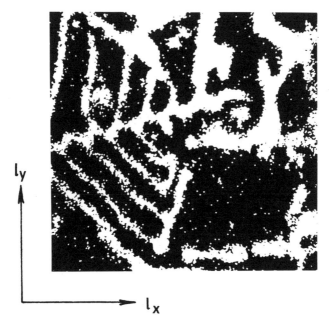

Figure 10.3. Qualitative planar distribution analysis: distribution image of Al on a surface of a lamellar Al–Si eutectic by EPMA

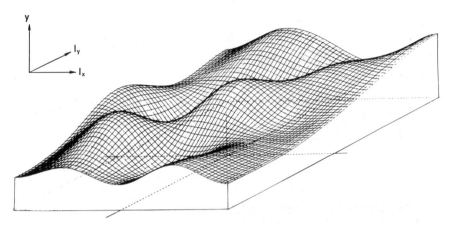

Figure 10.4. Quantitative planar element distribution $y = f(l_x, l_y)$ (concentration profile on a surface) of manganese in alloyed steel, determined by micro-spark OES (Danzer 1984)

where l_i is the diameter (cross section) of the entire sample under investigation, Δl_i the smallest resolvable lateral distance, given by the procedure (*lateral resolving limit*, diameter of measuring point), and i is the index of spatial direction (x, y, z).

Figure 10.6 shows that a decrease of the resolvable distance Δl_i is associated by an increase of resolving power A^* and therefore by an increase of information [see Eq. (10.4)]. Of course with a decrease of resolving power (e.g., by increasing a probe diameter), detailed lateral information is increasingly lost.

Plane resolving power A_p^* is written

$$A_p^* = \frac{p}{\Delta p} \tag{10.2}$$

where p is the plane (area) of the entire sample under investigation, $\Delta p = \Delta l_i \, \Delta l_j \approx (\Delta l_i)^2$ the smallest resolvable element of plane, the *plane resolving limit* ("pixel" according to Rüdenauer and Steiger 1984). The plane resolving power A_p^* is an important feature of all planar microprobe techniques like *electron probe micro analysis* (EPMA), *scanning Auger microprobe* (SAM), and *2D secondary ion mass spectrometry* (SIMS).

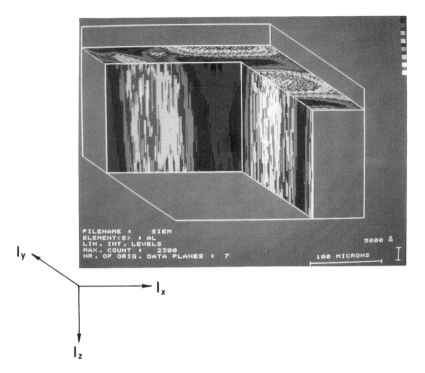

Figure 10.5. Quantitative 3D distribution analysis of Al on an integrated circuit ("pie section") (from Rüdenauer and Steiger 1986). Different concentrations are by the various grey steps.

Volume resolving power is written

$$A_v^* = \frac{v}{\Delta v} \qquad (10.3)$$

where v is the volume of the sample under investigation, $\Delta v = \Delta l_i \, \Delta l_j \, \Delta l_k \approx (\Delta l_i)^3$ the smallest resolvable volume element, *volume resolving limit* ("voxel,"; according to Rüdenauer and Steiger 1984, refers to the excitation volume in the sample, or interaction volume). The *ion microprobe* (SIMS) is a well-known method that directly yields three-dimensional analytical information using volume resolving power.

In the special case of Eq. (10.1) where $l_i = l_z$ is fixed perpendicularly to the surface of the sample, the *depth resolving power A_d^** =

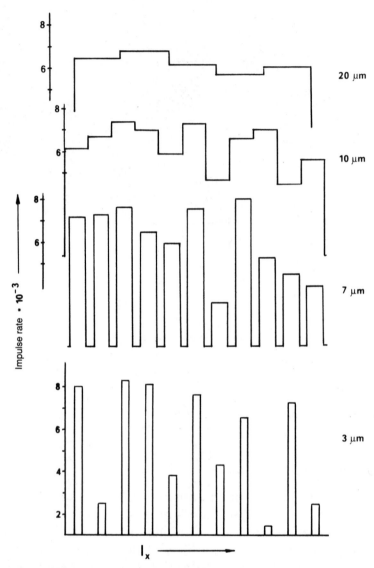

Figure 10.6. Distribution of Al on a lamellar eutectic: line scan across Fig. 10.3, analyzed by different lateral resolutions with different electron beam diameters of EPMA

$l_z/\Delta l_z$, which becomes important for depth profile and thin-film analysis.

10.2 POTENTIAL INFORMATION AMOUNT OF DISTRIBUTION-ANALYTICAL METHODS

For distribution-analytical procedures the potential (ideal) information amount M_{id} increases by the factor A^* in the case where all distinguishable points are analyzed. For each point information on maximum N components and their content is obtained as

$$M_{id} = A^* \cdot M_{id}(I) \qquad (10.4)$$

$M_{id}(I)$ is the potential (ideal) information amount of an analytical method with an integrating character that yields average analytical information about a given sample. For the application of this ideal information amount, the suppositions in Sections 6.1 and 7.3 must be fulfilled. From Eq. (7.24) we have the potential information amount

$$M_{id}(I) = M_p = N \cdot I_{max} = N \cdot 1b\, m_c \qquad (10.5)$$

where N is the analytical resolving power and m_c the number of discriminatable steps of concentration (concentration resolving power); see Eq. (6.9). For the potential (ideal) amount of information of distribution analyses M_{id}, we have

$$M_{id} = A^* \cdot N \cdot 1b\, m_c \qquad (10.6)$$

Referring to Table 10.1, where we compiled the analytical functions and the equations for the information amount of different procedures according to Fig. 10.1, we find that we can obtain the same expression for the maximum information amount for all the methods of distribution analysis. (1.0 through 1.3, 2.1 through 2.3, and so on). The reason is that the instruments and the parameters must be the same for point-, line-, and plane-distribution analysis to reach a certain resolving limit if the respective distribution analysis are realized by scanning procedures. In practice, the differences between local-, line-, and plane-distribution analyses are expressed in the information amount.

Table 10.1. Analytical Functions and Information Amounts
of the Distribution-Analytical Procedures and Problems Represented in Fig. 10.1

Distribution-Analytical Procedure	Analytical Function, $y, z =$	Potential (ideal) Information Amount of Procedure, $M_{id} =$	Practical Information Amount of Distribution-Analytical Result, $M_{d,r} =$
1.0 Local analysis (point analysis)	$\delta(l_x, l_y, l_z)$	$A_v \cdot M_{id}(I) = A_v \cdot N$ 1b m_c	$M(I) = n \cdot$ 1b m_c
1.1 Lateral distribution analysis (line scan; Fig. 10.2a)	$f(l_x)$	$A_v \cdot M_{id}(I) = A_v \cdot N$ 1b m_c	$a_x \cdot M(I)$
1.2 Regular plane distribution analysis (plane scan; Figs. 10.3 and 10.4)	$f(l_x, l_y)$	$A_v \cdot M_{id}(I) = A_v \cdot N$ 1b m_c	$a_x a_y \cdot M(I)$
1.2* Stochastic plane distribution analysis (simple testing of homogeneity)	—	$A_v \cdot M_{id}(I) = A_v \cdot N$ 1b m_c	$a_{xy} \cdot M(I)$
1.3 Volume distribution analysis (volume scan = combination of plane scan and sputtering, etching, etc.)	$f(l_x, l_y, l_z)$	$A_v \cdot M_{id}(I) = A_v \cdot N$ 1b m_c	$a_x a_y a_z \cdot M(I)$
1.3* Stochastic volume distribution analysis (stochastic testing of homogeneity in volume)	—	$A_v \cdot M_{id}(I) = A_v \cdot N$ 1b m_c	$a_{xyz} \cdot M(I)$
2.1 Integral lateral analysis (Fig. 10.2b)	$\delta(l_x, l_z)\int f(l_y)\,dl_y$	$A_p \cdot M_{id}(I)$	$M(I)$
2.2 Partial-integral plane analysis	$\delta(l_z)f(l_x)\int f(l_y)\,dl_y$	$A_p \cdot M_{id}(I)$	$a_x \cdot M(I)$
2.3 Partial-integral volume analysis	$f(l_x, l_z)\int f(l_y)\,dl_y$	$A_p \cdot M_{id}(I)$	$a_x a_z \cdot M(I)$
3.2 Integral plane analysis (surface and thin-film analysis)	$\delta(l_z)\int\int f(l_x, l_y)\,dl_x\,dl_y$	$A_d \cdot M_{id}(I)$	$M(I)$
3.3 Layer-integral volume analysis (layer-by-layer analysis)	$f(l_z)\int\int f(l_x, l_y)\,dl_x\,dl_y$	$A_d \cdot M_{id}(I)$	$a_z \cdot M(I)$
4.3 Average analysis	$\int\int\int f(l_x, l_y, l_z)\,dl_x\,dl_y\,dl_z$	$M_{id}(I)$	$M(I)$

Table 10.2. Estimated Values of the Geometrical Resolving Power A^* of Selected Distribution-Analytical Methods

Method	Geometrical Resolving Power A^*	
Microspark OES	$1 \cdots 10^2$	
Spark source mass spectroscopy	$10^2 \cdots 10^4$	
Laser OES, laser MS	$10^2 \cdots 10^4$	
Induced autoradiography	$10^4 \cdots 10^6$	
Electron probe microanalysis	$10^4 \cdots 10^6$	
Ion probe microanalysis (SIMS)	$10^4 \cdots 10^6$	
	$10^8 \cdots 10^{10}$	(A_v^*)
Electron microscopy	$10^8 \cdots 10^{11}$	
Field ion microscopy—atom probe	$10^{12} \cdots 10^{14}$	

Source: Danzer (1974).
Note: A_p^* related to $p = 1$ mm^2, A_v^* to $v = 0.1$ mm^3.

Table 10.1 also indicates that the information amount of the procedures decreases with increasing integration (i.e., from local analysis to integral line, integral plane and to average analysis). In each case the regions whose compositions are sampled are those whose sizes differ considerably (see 1.0, 2.1, 3.2, and 4.3 in Fig. 10.1). But the nature of the analytical information is the same in all four cases: We are informed about the presence of maximum N components and their contents (1b m). Roughly calculated values for the geometrical resolving power A^* of some important distribution-analytical methods are given in Table 10.2.

The expressions for the information amounts according to Eqs. (10.4), (10.6), and in Table 10.2 do not take into account the loss of precision of the results with increasing geometrical resolving power nor the errors in adjusting the measuring points. Furthermore the independence of all the A^* measuring points must be presupposed for an estimation of the ideal information amount. The influence of the real conditions, correlations, and interactions on this relationship is explained by Ehrlich, Danzer, and Liebich (1979).

Considering the large increase of information amount by the geometrical resolving power, we will always have a much higher information amount of distribution-analytical investigations than of average analyses, and likewise for cases where the precision of the quantitative determination is drastically decreased. If we introduce

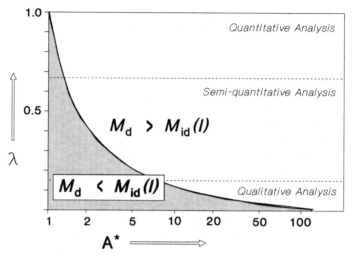

Figure 10.7. Comparison of distribution-analytical and average-analytical information amounts

λ as a measure of the loss of precision, we can express the relation between the information amount of distribution analysis M_d and that of average analysis $M_{id}(I)$ by

$$M_d = \lambda \cdot A^* \cdot M_{id}(I) \tag{10.7}$$

The values of λ can be assumed as follows:

$1.00 > \lambda > 0.67$ if no considerable loss of precision is observed; for each area under investigation still quantitative results are obtained

$0.67 > \lambda > 0.15$ if instead of quantitative only semiquantitative results are obtainable

$\lambda \leq 0.15$ if only qualitative information on the sample areas are accessible

The curve in Fig. 10.7 connects the pairs of values of λ and A^* for which $M_d = M_{id}(I)$ is valid. The figure shows that in cases where quantitative results can still be obtained, the information gain is registered with $A \geq 2$ for the distribution analysis compared with the average analysis. If only qualitative analysis can be obtained with $A \geq 7$, a higher information amount is present for distribution than for average analysis. We know that this is a plausible case because the most precise and accurate analysis will depreciate if fluctuations of contents are detected; for example, see Fig. 10.3.

10.3   INFORMATION AMOUNT OF DISTRIBUTION-ANALYTICAL RESULTS

Another limiting case is given by the real information amount $M_{d,r}$ of a given distribution-analytical result (index d for distribution analysis, r for real):

$$M_{d,r} = a \cdot M(I) = a \cdot n \cdot 1b\, m_c \qquad (10.8)$$

Equation (10.8) is determined, in practice, by the geometrical resolving power a. Therefore the information amount of a distribution-analytical result is a times greater than that of integral analysis $M(I)$, which is given by the number n of analyzed components and the maximum information content $1b\,m$ of each component [Eq. (6.9)].

The number a of measuring points (e.g., spark spots or laser craters in optical emission spectroscopy or mass spectroscopy) is determined by the given problem, such as testing for homogeneity. However, the real information amount $M_{d,r}$ is lower than the ideal M_{id} because $a < A^*$ and $n < N$.

In the right column of Table 10.1 the real information amounts are listed for the different procedures of Fig. 10.1. We had seen that the potential information amount M_{id} remains constant for all cases without integration (cases 1.0 through 1.3, 1.2*, 1.3*) because of a constant number of distinguishable points from a complete inspection of the sample. On the other hand, the real information amount $M_{d,r}$ increases from the point- to the complete volume-distribution analysis, since ever more points, lines, and planes are inspected (cases 1.0 through 1.3, 2.1 through 2.3, and 3.2 through 3.3).

Similarly the real information amounts of analyses yielding only one concentration value for a volume element, namely local analysis, integral line and plane analyses, and average (total integral) analysis, are equal (cases 1.0, 2.1, 3.2, 4.3). This also contrasts with the ideal information amounts of the procedures.

We assume that the measuring points are independent as in the cases of ideal information amount of procedures and real information amount of distribution analyses. But in distribution-analytical practice the measuring points may be mutually dependent either because the extended points touch one another or because they are scanned successively. In addition geometrical resolving power and precision are interrelated in proportion to the number of discriminatory steps. Details on these relationships can be found in Ehrlich, Danzer, Liebich (1979).

10.4   SIMULTANEOUS DETERMINATION OF CONTENT AND SAMPLE POSITION

In all the cases of distribution analysis treated above, the measuring strategy used mostly consists in the determination of concentrations c at well-defined, error-free adjusted subsample positions. Then the a priori known positions do not contribute to the information amount of the analytical result.

But from a more general point of view, the positioning of subsamples can be realized only with an uncertainty Δl_i of length l_i $(i = x, y, z)$ in an expected position. Therefore the complete data set $\{c, x, y, z\}$ obtained for a single (index s) measuring point and one analytical component yields the ideal information amount

$$M_{s,\,id} = 1b\, m_c + 1b\, m_x + 1b\, m_y + 1b\, m_z \qquad (10.9)$$

where m_c represents the discriminatable steps of concentration and $m_i = l_i/\Delta l_i$ $(i = x, y, z)$ are the discrimatory positions. Obviously each dimension, namely the concentration and spatial coordinates, contribute equally to the information amount.

As mentioned above, Eq. (10.9) can be multiplied by a factor N (potential) or n (real) analytical components and by a factor A^*

(ideal) or a (real) measuring points. Marginal cases of Eq. (10.9) are those involving exact positioning and exactly preselected concentrations. In the first case, given $\Delta l_i = l_i$, then $m_i = 1$ for each spatial coordinate, leading to Eqs. (10.6) and (10.8) for ideal and real information amounts, respectively. In the second case, when the related coordinates are searched, with $m_c = 1$, the term 1b m_c equals zero, and the information amount is expressed only by the remaining "geometrical terms" 1b m_i (e.g., for one component and two spatial coordinates):

$$M_{id, c=const.} = 1b\, m_x + 1b\, m_y \qquad (10.10)$$

In practice, Eq. (10.10) would appear in a plane as an "iso-concentration" line. Figure 10.8 shows a collection of iso-concentration lines obtained from a surface analysis for the element distribution of Fig. 10.4. Images of this type are useful in assessing the intended changes of concentration (e.g., by diffusion) or for the representation of the homogeneity of solid samples.

10.5 TESTING FOR CHEMICAL HOMOGENEITY

The investigation of homogeneity in solid samples is a curtailed version of the distribution analysis (Ehrlich and Mai 1982; Danzer and Ehrlich 1984). Chemical homogeneity is a relative property of a material depending on the amount of subsample under investigation (and therefore on its area or volume) and on the tolerated concentration fluctuations given by the application of the material. The fundamentals of homogeneity tests are described in Danzer et al. (1979) and Danzer and Ehrlich (1984).

The special features of homogeneity testing are the following:

1. Simplified distribution-analytical investigation of a given area or volume of a solid carried out by measurements at stochastically arranged points (see Fig. 10.1, procedure 1.2*) or regularly in r rows and c columns with $a = a_r a_c$ (Fig. 10.1, procedure 1.2), where in general $a_r a_c \ll a_x a_y$.

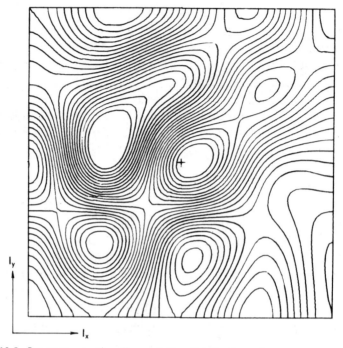

Figure 10.8. Iso-concentration lines of the distribution of manganese on alloyed steel as another kind of concentration profile on a surface; determined by microspark OES (Danzer 1984); see Fig. 10.4

2. Testing of significant differences in a chemical composition by statistical methods dependent on the characterization of inhomogeneity in the given case (the type of concentration fluctuation); different statistical models can be found in Danzer et al. (1977, 1979, 1984, 1985), Ehrlich and Mai (1982), Inczedy (1982), Ehrlich and Kluge (1989), Liebich et al. (1989), and Parczewski, Danzer, and Singer (1986).

Essentially the information from homogeneity testing is obtained in two parts: from a distribution analysis with the result $M_{d,r}$ [Eq. (10.8)], which depends on the number of measuring points and its geometrical arrangement, and from the information content I_h of a statistical analysis for homogeneity. For the latter the expression

$$I_h = (a - 1)\mathrm{lb}\, h \qquad (10.11)$$

was proposed by Danzer, Schubert, and Liebich (1991). The information content of the statistical result depends on the number of degrees of freedom, $a - 1$, and on h that describes how weighty the decision about homogeneity or inhomogeneity is.

The decision can be quantified by an *index of homogeneity H* (defined by Danzer 1984; see also Singer and Danzer 1984). More convenient is a definition using α-values (risk-of-error values) given by modern statistical software packages as results of statistical tests. The *H*-value is defined by the ratio of a given tolerable risk of error α_{crit} and the risk α estimated by the test statistics:

$$H_a = \frac{\alpha_{\mathrm{crit}}}{\alpha} \qquad (10.12)$$

For instance, for Fischer's F-statistics, α is an inverse function F^{-1} to

$$\hat{F}(\alpha, f_t, f_p) = \frac{s_t^2}{s_p^2} \qquad (10.13)$$

where s_t and s_p are the total error and the error of the analytical procedure, respectively, and f_t and f_p the corresponding degrees of freedom.

For the purpose of information-theoretic characterization it is advantageous to use a *sharpness-of-homogeneity decision function*

$$h = \begin{cases} H_\alpha & \text{if } H_\alpha \geq 1 \\ H_\alpha^{-1} & \text{if } H_\alpha < 1 \end{cases} \qquad (10.14)$$

(as defined by Danzer, Schubert, and Liebich 1991) because the more H differs from 1, the stronger is the decision against homogeneity on the one side and for acceptance of homogeneity on the other. Likewise the information content will be the greater, the more H diverges from 1, no matter if they be higher or lower values as Fig. 10.9 shows.

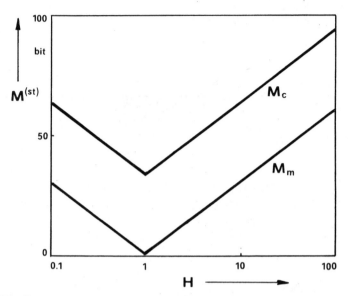

Figure 10.9. Dependence of information amount M on the index of homogeneity for one component and for a stochastical arrangement of measuring points, where M_c = use of concentration values, M_m = use of measuring values; number of points $a = 10$, precision $m_c = 10$

Decisions about homogeneity can be made also on the basis of measured values (e.g., X-ray fluorescence intensities) instead of concentrations. Since the measured values are not the final aim of the chemical analysis, the relevant information consists then only of the statement of homogeneity or inhomogeneity.

Table 10.3 gives an example of different investigations of homogeneity. From the table it is clear that compared with homogeneity obtained by a regular arrangement of measuring points, an evaluation of concentration values yields the highest information content. One reason is that not only is a general assessment of the total homogeneity h_t possible, but from the multiple dispersion decomposition (two-way analysis of variance, as outlined by Danzer and Marx 1979) also preferred directions of inhomogeneity (gradients of concentration) are obtainable from homogeneity testing within rows h_r and within columns h_c.

However, the smallest information amounts result from tests that stochastically arrange measuring values $M_{h,m}^{(st)}$. As shown in

Table 10.3. Information Contents for Different Procedures of Distribution Analytical Techniques (Related to One Component in Each Case)

	Testing of Homogeneity by	
	Measured Value	Concentration Value
Stochastically arranged measuring points	$M_{h,m}^{(st)} = (a - 1)\text{lb } h$ $\qquad (10.11)$	$M_{h,c}^{(st)} = a \text{ lb } m_c$ $+(a - 1)\text{lb } h$ $= M_{d,r} + M_{h,m}(st) \qquad (10.15)$
Regularly arranged measuring points	$M_{h,m}^{(rg)} = (a_r a_c - 1)\text{lb } h_t$ $+(a_r - 1)\text{lb } h_c$ $+(a_c - 1)\text{lb } h_r \quad (10.16)$	$M_{h,c}^{(rg)} = a_r a_c \text{ lb } m_c + M_{h,m}^{(rg)}$ $= M_{d,r} + M_{h,m}^{(rg)} \qquad (10.17)$

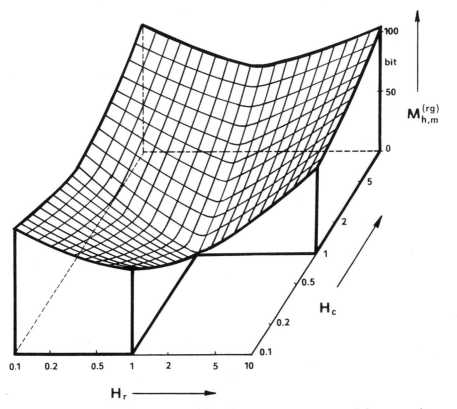

Figure 10.10. Information amount $M_{h,m}$ for one component and for a regular arrangement of measuring points ($a_r = 5$, $a_c = 5$) which depend on the index of homogeneity H_c and H_r (H_t = the average of H_c and H_r)

Fig. 10.10, such information depends on the index of homogeneity for two directions of space in the case of regularly arranged measuring points.

The preceding remarks refer to the homogeneity of one element (univariate testing). There are two possible ways of investigating several elements. In practice, the homogeneity for n components is usually tested by a multiple procedure. The total information amount then represents the sum of the information amounts for each component:

$$M_h(n) = \sum_{i=1}^{n} (M_h)_i \qquad (10.18)$$

An alternate way to investigate the homogeneity of a sample is by multivariate data analysis, especially with pattern recognition methods (see Danzer and Singer 1984; Ehrlich, Danzer, and Kluge 1985).

In both methods the n elements are considered simultaneously according to their connections and interactions. Whereas by Eq. (10.18) the multiple case refers either to concentration values (distribution analysis) or only to decisions about homogeneity, the multivariate case always concerns a general statement of homogeneity. The homogeneity information is given as

$$M_{mv} = n(a - 1)1b\, h_{mv} \qquad (10.19)$$

for n components and a measuring points.

A wide range of analytical and chemometrical methods is available for testing the homogeneity of solids. Information theory makes it possible to characterize the methods and results of these special (curtailed) variants of distribution analysis according to their sampling state (arrangement of measuring points), procedures of evaluation (using concentrations or measuring values), and univariate, multiple, or multivariate characteristics.

10.6   PRACTICAL APPLICATIONS

In practice, the information amount of distribution analyses depends on both the parameters of the analytical procedure and the characteristics of the material under investigation. This will be illustrated by a general example on the real information amount of distribution-analytical procedures in the case of multiphase materials (Schubert 1989; Danzer, Schubert, and Liebich 1991).

In general, the stochastic properties of a multiphase solid material can be given by the Gaussian probability density function

$$p(c) = \frac{1}{\sqrt{(\pi \bar{p} / \Delta p)}} \exp\left[-\frac{(c - \bar{c})^2}{2 \bar{p} / \Delta p} \right] \qquad (10.20)$$

where c is the average concentration in a defined homogeneous plane element $\bar{p}$, and $\bar{c}$ is the average concentration of the total analyzed plane p where $p \gg \bar{p}$.

To perform an instructive treatment, we introduce sufficiently small thresholds with the distance $2\Delta c$ (confidence interval), which allows the assignment of the measured c-values with a small error probability. To this end we replace the probability distribution $p(c)$ of Eq. (10.20) by

$$w(c) = \begin{cases} k \cdot p(c) & \text{for } |c - \bar{c}| \leq \Delta c \\ 0 & \text{for } |c - \bar{c}| > \Delta c \end{cases} \qquad (10.21)$$

with k as the normalization factor. From

$$\Delta c = \sqrt{\frac{\bar{p}}{(\Delta p)}} \qquad (10.22)$$

it follows that the number of discriminatory steps

$$m_c = \frac{c_{\max} - c_{\min}}{2\Delta c} = \frac{c_{\max} - c_{\min}}{\sqrt{(2\bar{p}/\Delta p)}}$$

$$= \frac{c_{\max} - c_{\min}}{\sqrt{(2\bar{l}/\Delta l)}} \qquad (10.23)$$

where $\bar{l} \approx \sqrt{\bar{p}}$ is the average diameter of the homogeneous plane elements and $\Delta l \approx \sqrt{\Delta p}$ the lateral resolving of the analytical procedure.

In actual cases it is more sensible to assume that the errors due to the material Δc_{mat} and due to the measurement method Δc_{meas} do not correlate. Therefore Eq. (10.23) for the number of discriminatory steps should be modified as

$$m_c = \frac{c_{max} - c_{min}}{2\sqrt{(\Delta c_{mat}^2 + \Delta c_{meas}^2)}} \qquad (10.24)$$

From Eq. (10.23) we have, using the basic relation $A_p = p/\Delta p$ of Eq. (10.2),

$$\Delta p = \frac{2\bar{p}m_c^2}{(c_{max} - c_{min})^2} \qquad (10.25)$$

and

$$A_p = \frac{p(c_{max} - c_{min})^2}{2\bar{p}m_c^2} \qquad (10.26)$$

Clearly both A_p and m_c depend on the concentration interval and on geometrical factors of material $\bar{p}$ and analytical procedure Δp. Both quantities show interdependence, since A is reciprocally proportional to m_c^2. Therefore the real information amount is given by

$$M = \frac{p(c_{max} - c_{min})^2}{2\bar{p}m_c^2} \, 1b \, m_c \qquad (10.27)$$

or

$$M = \frac{p}{\Delta p} \, 1b \left[(c_{max} - c_{min}) \sqrt{\frac{\Delta p}{2\bar{p}}} \right] \qquad (10.28)$$

Equations (10.27) and (10.28) show that the information stored in a material is *finite*. As can be readily seen, the maximum value of the function $(1b \, m_c)/m_c^2$ dependent on the integers m_c is attained for $m_c = 2$. In practice this case of binary coding of element concentration is given by image-analytical techniques like the electron microprobe (EMPA; see Fig. 10.3).

An optimum area p_{opt} can be found for which a maximum information for image-analytical procedures results in the case of binary coded element information ($m_c = 2$):

$$p_{opt}^{(2)} = \frac{8\bar{p}}{\left(c_{max} - c_{min}\right)^2} \qquad (10.29)$$

We know from our distribution-analytical experiences that the area under investigation must be larger as the homogeneous plane elements (e.g., phases) $\bar{p}$ of the sample become more extensive and the difference between the concentrations in the phases smaller.

A quasi-optimum area may be assumed when there are more than two discriminatory concentration steps $m_c > 2$, as in the case of image-analytical procedures using color-coded techniques,

$$p_{opt} = \frac{2\bar{p}m_c^2}{\left(c_{max} - c_{min}\right)^2 \mathrm{1b}\, m_c} \qquad (10.30)$$

The result is naturally larger than that in the dual-coded case of Eq. (10.29).

Finally, we turn to some examples of the information amount of distribution analyses. To begin with, we consider the EPMA element imaging of a lamellar Al-Si eutect. Figure 10.3 gives a cross section of $l = 250$ μm and consequently an area of $p = 62,500$ μm^2. With a diameter of the electron beam $\Delta l \approx 5$ μm and $\Delta p \approx 25$ μm^2, we have by Eq. (10.2) a plane resolving power of $A_p \approx 2500$ and by Eq. (10.6), with $N = 1$ and $m_c = 2$, an information amount $M \approx 2500$ bit.

Next is the case of the three-dimensional distribution analysis of Al in an integrated circuit; see Fig. 10.5 (Rüdenauer and Steiger 1986), the geometrical parameters are $l_x \approx l_y \approx 250$ μm, $l_z \approx 4.25$ μm, $\Delta l_x \approx \Delta l_y \approx 5$ μm, $\Delta l_z \approx 0.5$ μm, and therefore $v \approx 265,625$ μm^3 and $\Delta v \approx 12.5$ μm^3. The volume resolving power $A_v \approx 21,250$ by Eq. (10.3), and the information amount by Eq. (10.6), with $N = 1$ and $m_c = 10$ different grey steps, is $M \approx 70,593$ bit.

The information amount of testing for chemical homogeneity are illustrated in investigations by Danzer and Marx (1979) with regard to the distribution of chromium and manganese in steel. Their evaluation of homogeneity was carried out on the basis of regularly arranged measuring points (5 × 5) by a two-way analysis of variance. The quantities used for the calculation are

Cr: $c_{max} - c_{min} = \bar{c} = 1.83$ wt-%, $\Delta c = 0.008$ wt-% $H_t = 0.5875$, $h_t = 1.7021$; $H_r = 0.2313$, $h_r = 4.3234$, $H_c = 0.9431$, $h_c = 1.0603$[1] (homogeneous distribution of chromium)

Mn: $c_{max} - c_{min} = \bar{c} = 1.68$ wt-%, $\Delta c = 0.0177$ wt-% $H_t = 1.2583$, $H_r = 1.2456$, $H_c = 1.3238$ (inhomogeneous distribution of manganese; see Figs. 10.4 and 10.8)

where the indexes of homogeneity (according to Danzer 1984) are total (in)homogeneity H_t, (in)homogeneity in direction of rows H_r, and columns H_c.

From Eqs. (10.16) and (10.17) the measured and the concentration values yield the following information content:

$$M(\text{Cr})_{h,m}^{(rg)} = 24 \, lb \, 1.702 + 4 \, lb \, 4.323 + 4 \, lb \, 1.06 = 27.2 \text{ bit}$$

$$M(\text{Cr})_{h,c}^{(rg)} = 25 \, lb \frac{1.83}{0.016} + M(\text{Cr})_{h,m}^{(rg)}$$

$$= 170.9 \qquad + 27.2 \qquad\qquad = 198.1 \text{ bit}$$

$$M(\text{Mn})_{h,m}^{(rg)} = 24 \, lb \, 1.258 + 4 \, lb \, 1.246 + 4 \, lb \, 1.324 = 10.8 \text{ bit}$$

$$M(\text{Mn})_{h,c}^{(rg)} = 25 \, lb \frac{1.68}{0.0354} + M(\text{Mn})_{h,m}^{(rg)}$$

$$= 139.2 \qquad + 10.8 \qquad\qquad = 150.0 \text{ bit}$$

As we can see, the information content calculated from measured values $M_{h,m}$ is independent of whether the finding is homogeneous or inhomogeneous; it depends only on sharp decisions. The information content based on concentration values $M_{h,c}$ is much higher

[1] The index of homogeneity H is defined by Danzer (1984) and Danzer and Singer (1984), and h according to Eq. (10.14).

than $M_{h,m}$ and depends on the expectation value, from Eqs. (6.7) and (6.8), and precision (as discussed in Chapter 6).

For information relating to both elements and the material in a more general way, we can use the univariate calculation of Eq. (10.18) to get $M(\mathrm{Cr},\mathrm{Mn})_{h,m}^{(\mathrm{rg})} = 38.0$ bit and $M(\mathrm{Cr},\mathrm{Mn})_{h,c}^{(\mathrm{rg})} = 348.1$ bit. For an average analysis of both elements in the steel with comparable precision, we can obtain $M(\mathrm{I})_{\mathrm{Cr}+\mathrm{Mn}} = 1\mathrm{b}\ (1.83/0.016) + 1\mathrm{b}(1.68/0.0354) = (6.84 + 5.57)$ bit $= 12.4$ bit. With spatially resolved information and confirmation of homogeneity, the information increases about the 28-fold.

REFERENCES

Danzer, K. 1974. *Z. Chem.* **14**, 73.

Danzer, K. 1984. *Spectrochim. Acta* **39B**, 949.

Danzer, K., Doerffel, K., Ehrhardt, H., Geissler, H., Ehrlich, G., and Gadow, P. 1979. *Anal. Chim. Acta* **105**, 1.

Danzer, K., and Ehrlich, G. 1984. Tagungsber. Techn. Hochsch. Karl-Marx-Stadt: 4. Tagung Festkörperanalytik, vol. 2, p. 547.

Danzer, K., and Küchler, L. 1977. *Talanta* **24**, 561.

Danzer, K., and Marx, G. 1979. *Anal. Chim. Acta* **110**, 145.

Danzer, K., Schubert, M., and Liebich, V. 1991. *Fresenius J. Anal. Chem.* **341**, 511.

Danzer, K., and Singer, R. 1985. *Mikrochim. Acta* [Vienna] **I**, 219.

Danzer, K., Than, E., Molch, D., and Küchler, L. 1988. *Analytik— Systematischer Überblick*. 2d ed., Akademische Verlagsgesellschaft Geest & Portig, Leipzig.

Ehrlich, G., Danzer, K., and Kluge, W. 1985. Proc. Sixth International Symposium on High-Purity Materials in Science and Technology, Dresden. Vol. II: Characterization, p. 59.

Ehrlich, G., Danzer, K., and Liebich, V. 1979. Tagungsber. Techn. Hochsch. Karl-Marx-Stadt: 2. Tagung Festkörperanalytik, vol. 1, p. 69.

Ehrlich, G., and Kluge, W. 1989. *Mikrochim. Acta* [Vienna] **I**, 145.

Ehrlich, G., and Mai, H. 1982. Tagungsber. Techn. Hochsch. Karl-Marx-Stadt: 3. Tagung Festkörperanalytik, vol. 1, p. 97.

Inczedy, J. 1982. *Talanta* **29**, 643.

Liebich, V., Ehrlich, G., Stahlberg, U., and Kluge, W. 1989. *Fresenius Z. Anal. Chem.* **335**, 945.

Parczewski, A., Danzer, K., and Singer, R. 1986. *Anal. Chim. Acta* **191**, 461.

Rüdenauer, F. G., and Steiger, W. 1984. *Surface Interface Anal.* **6**, 132.

Rüdenauer, F. G., and Steiger, W. 1986. Wiss. Beitr. Friedrich Schiller University, Jena. 3. Tagung Computereinsatz in der Analytik—COMPANA '85, p. 115.

Schubert, M. 1989. Private communication.

Singer, R., and Danzer, K. 1984. *Z. Chem.* **24**, 339.

STRUCTURE ANALYSIS

Structure analysis deals with the arrangement and bonding of elementary units in molecules, macromolecules, or crystals and their symmetry and geometry. According to Danzer et al. (1976) structure analysis can be considered a distribution analysis in atomic dimensions characterized by the function

$$z = f(l_x, l_y, l_z)_{y=1\text{su}} \tag{11.1}$$

where y represents one structural unit and l_x, l_y, and l_z are in the nanometer range. In practice, structure analysis is regarded as a separate field because of its theoretical and methodological particularities.

Structures are characterized qualitatively and quantitatively by *structure matrices*[1]

$$\underline{\mathbf{ST}} = \begin{pmatrix} z_1 & z_2 & z_3 & \cdots & z_N \\ 0 & b_{12} & b_{13} & \cdots & b_{1N} \\ b_{21} & 0 & b_{23} & \cdots & b_{2N} \\ \vdots & \vdots & \vdots & & \vdots \\ b_{N1} & b_{N2} & b_{N3} & \cdots & 0 \end{pmatrix} \tag{11.2}$$

Structure matrices describe mathematically the kind and degree of connection between structural units, where b_{ij} characterizes the connection (bonding) between two structure units z_i and z_j. Though for the sake of visualization, instead of the structure matrices themselves, their graphs are commonly used as structural formulas,

---

[1]Since the symbol $\underline{\mathbf{S}}$ refers to the variance/covariance matrix, the unusual double symbol $\underline{\mathbf{ST}}$ is used for the structure matrix.

structure matrices embody the complete information on the structure of molecules, macromolecules, and crystals.

By analogy with elemental analysis, it is reasonable to make a distinction between *qualitative* and *quantitative* structure analysis. Qualitative structure matrices $\underline{\mathbf{ST}} = (b_{ij})$ describe the bonding of individual structural units $z_1, z_2, \ldots, z_N$ (atoms, structural groups, etc.) in a qualitative way, where b_{ij} can be ordinal or natural (cardinal) numbers, for example,

$$b_{ij} = \{0, 1, 2, \ldots\} \qquad (11.3)$$

or binary numbers, for example,

$$b_{ij} = \{0, 1\} \quad \text{or} \quad b_{ij}\{0, b\}$$

The quantity b_{ij} assumes the value b or 0, depending on whether the structural elements z_i and z_j are connected with each other. Figure 11.1 illustrates these facts.

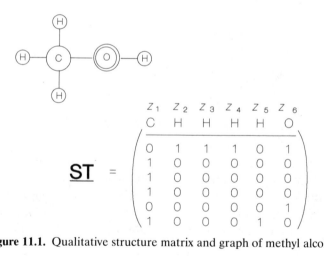

Figure 11.1. Qualitative structure matrix and graph of methyl alcohol

In contrast, *quantitative structure matrices*

$$
\underline{\underline{ST}} = \begin{pmatrix}
z_1 & z_2 & z_3 & \cdots & z_N \\
\hline
0 & \underline{b}_{12} & \underline{b}_{13} & \cdots & \underline{b}_{1N} \\
\underline{b}_{21} & 0 & \underline{b}_{23} & \cdots & \underline{b}_{2N} \\
\vdots & \vdots & \vdots & & \vdots \\
\underline{b}_{N1} & \underline{b}_{N2} & \underline{b}_{N3} & \cdots & 0
\end{pmatrix}
\tag{11.4}
$$

contain bonding vectors

$$
\begin{aligned}
\underline{b}_{ij} &= \underline{x} b_x + \underline{y} b_y + \underline{z} b_z \\
&= b_{ij} \left(\underline{x} \cos \alpha + \underline{y} \cos \beta + \underline{z} \cos \gamma \right)
\end{aligned}
\tag{11.5}
$$

with bond length or lattice spacing

$$
b_{ij} = \sqrt{\underline{b}_x^2 + \underline{b}_y^2 + \underline{b}_z^2}
\tag{11.6}
$$

and bond angles

$$
\alpha = \cos^{-1} \frac{\underline{b}_x}{b_{ij}}
$$

$$
\beta = \cos^{-1} \frac{\underline{b}_y}{b_{ij}}
\tag{11.7}
$$

$$
\gamma = \cos^{-1} \frac{\underline{b}_z}{b_{ij}}
$$

by which the bonding of two structure elements z_i and z_j is described quantitatively, as illustrated in Fig. 11.2. The complete structure analysis for molecules or crystals can be consolidated schematically into the steps listed in Table 11.1.

While elemental analyses are normally concerned with mixtures of substances, structure analyses are carried out almost exclusively on pure substances after separation and purification. The first two

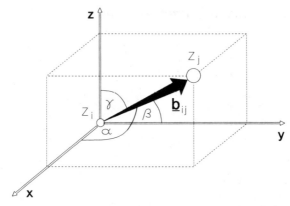

Figure 11.2. Bonding vector $\underline{b}_{ij}$ between two structure elements z_i and z_j

steps given in Table 11.1 identify the compound, which is a problem of elemental analysis (see Chapter 4).

Element-analytical investigations, for molecules such as in classical organic elementary analysis and the determination of molar weight or mass spectroscopy, are used for the identification of substances. As result a total formula is obtained and therefore an unambiguous composition. The latter is the basis for structure analysis in a narrower sense, starting with the third step in Table 11.1.

Table 11.1. Steps of Structure Analysis of Molecules and Crystals

Partial Step in Determination of	Information on	
	Molecular Structure	Crystal Structure
Elemental composition	Proportional formula	Proportional formula
Number of structure elements	Total formula	Total formula
Qualitative structure matrix	Constitution	Type of lattice, crystal system
Symmetry	Configuration, conformation	Crystal class, point group
Quantitative structure matrix	Geometry of the molecule (bond lengths and angles)	Lattice spacings of the unit cell

11.1 INFORMATION AMOUNT OF QUALITATIVE STRUCTURE ANALYSIS

Information on the bonding of components, represented by constitutional formulas or structure matrices $\underline{ST}$, are qualitative in nature, since they do generally not explain the stereometric structure of molecules. Qualitative structure matrices, $\underline{ST} = (b_{ij})$, are obtained if the composition and complete formula are known. The number of theoretically possible bonds of N structure elements is

$$a_N = \frac{N(N-1)}{2} \tag{11.8}$$

since any $b_{ii} = 0$ and $b_{ij} = b_{ji}$. The mathematical variety of possible bonds is limited to valence-theoretic relationships and rules.

The information amount for establishing the qualitative structure matrix is written

$$M(\underline{ST}) = \sum_{k=1}^{a_N} \sum_{l=1}^{q} P(b_l)_k \, \mathrm{lb}\big[P(b_1)_k \big]^{-1} \tag{11.9}$$

where $P(b_l)_k$ is the probability of certain bonding states ($l = 0, 1, 2, \ldots q$); for example, 1 = single bond, 2 = double bond, 3 = triple bond, 4 = aromatic bonding state.

In practice, spectroscopic methods above all provide information about structure groups (of which molecules, macromolecules, molecular and ionic crystals are composed) as well as about their bonding. In information-analytical terms this means a reduction of the original structure matrix just as the constitutional formulas or crystal structure representations undergo simplification. In particular, NMR–, UV/VIS–, IR–, Raman–, and mass spectroscopy, frequently in combination, are used to investigate the qualitative structure of molecules and crystals (as mentioned in Chapter 4).

Questions of *configuration* and *conformation* are generally answered by symmetry investigations and are beyond the scope of qualitative structural statements.

The qualitative structure matrix differs for different configurations of stereo-isomers, as shown by the example of two forms of

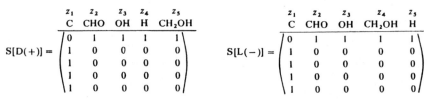

$$S[D(+)] = \begin{array}{ccccc} z_1 & z_2 & z_3 & z_4 & z_5 \\ C & CHO & OH & H & CH_2OH \\ \begin{pmatrix} 0 & 1 & 1 & 1 & 1 \\ 1 & 0 & 0 & 0 & 0 \\ 1 & 0 & 0 & 0 & 0 \\ 1 & 0 & 0 & 0 & 0 \\ 1 & 0 & 0 & 0 & 0 \end{pmatrix} \end{array} \qquad S[L(-)] = \begin{array}{ccccc} z_1 & z_2 & z_3 & z_4 & z_5 \\ C & CHO & OH & CH_2OH & H \\ \begin{pmatrix} 0 & 1 & 1 & 1 & 1 \\ 1 & 0 & 0 & 0 & 0 \\ 1 & 0 & 0 & 0 & 0 \\ 1 & 0 & 0 & 0 & 0 \\ 1 & 0 & 0 & 0 & 0 \end{pmatrix} \end{array}$$

Figure 11.3. Configurations of glyceraldehyde and their qualitative structure matrices

glyceraldehyde in Fig. 11.3, though the conformers of right-handed and left-handed quartz crystals cannot be so distinguished. The following holds, for example, for the two conformers of cyclohexane (Fig. 11.4):

$$\underline{ST}(\text{chair form}) = \underline{ST}(\text{boat form})$$

The differences can only be seen in the permutation groups, which are D_{3d} and C_{2v}, respectively.

Semiquantitative structure matrices are suited for different configurations or conformations, in which the bonding vectors are not given by their magnitude or angle, but by their direction (Danzer and Marx, 1979a, 1979b), as can be seen in Fig. 11.4. The information amount of such semiquantitative structural information increases with respect to the qualitative structure matrix by

$$M_{\text{sym}} = \sum_{r=1}^{u} P(\pi_r) \, 1b\left[P(\pi_r) \right]^{-1} \qquad (11.10)$$

where $P(\pi_r)$ is the probability of the occurrence of certain permu-

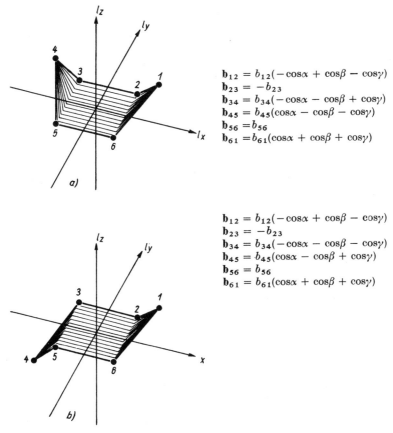

$$b_{12} = b_{12}(-\cos\alpha + \cos\beta - \cos\gamma)$$
$$b_{23} = -b_{23}$$
$$b_{34} = b_{34}(-\cos\alpha - \cos\beta + \cos\gamma)$$
$$b_{45} = b_{45}(\cos\alpha - \cos\beta - \cos\gamma)$$
$$b_{56} = b_{56}$$
$$b_{61} = b_{61}(\cos\alpha + \cos\beta + \cos\gamma)$$

$$b_{12} = b_{12}(-\cos\alpha + \cos\beta - \cos\gamma)$$
$$b_{23} = -b_{23}$$
$$b_{34} = b_{34}(-\cos\alpha - \cos\beta - \cos\gamma)$$
$$b_{45} = b_{45}(\cos\alpha - \cos\beta + \cos\gamma)$$
$$b_{56} = b_{56}$$
$$b_{61} = b_{61}(\cos\alpha + \cos\beta + \cos\gamma)$$

Figure 11.4. The structure matrice for boat form (a) and chair form (b) and their respective conformations of cyclohexane and bonding vectors

tation groups or certain symmetries. Finally, we have

$$M(\underline{ST})_{sq} = M(\underline{ST}) + M_{sym}$$

$$= \sum_{k=1}^{a_N} \sum_{l=1}^{q} P(b_l)_k \, \mathrm{lb}\big[P(b_l)_k\big]^{-1} + \sum_{r=1}^{u} P(\pi_r)\mathrm{lb}\big[P(\pi_r)\big]^{-1}$$

$$(11.11)$$

for the information amount of semiquantitative structure matrices.

11.2   INFORMATION AMOUNT OF QUANTITATIVE
STRUCTURE ANALYSIS

Quantitative structural information includes complete data about the location of all structural units in space, that is, about all interatomic distances and bonding angles. Accordingly the quantitative structure of a molecule or a unit cell is fully described by the *quantitative structure matrix*, $\underline{\underline{ST}} = (\underline{\mathbf{b}}_{ij})$, according to Eq. (11.4) with the bonding vectors $\underline{\mathbf{b}}_{ij}$ by Eq. (11.5).

The amount of quantitative structural information results from the expectation range, $\mathbf{\Delta}_{ex}\underline{\mathbf{b}}_{ij}$, of the determined vectors, $\underline{\mathbf{b}}_{ij} = (b_x, b_y, b_z)$, of measured values, b_i, required for the complete quantitative structure matrix and the remaining uncertainty of the measured values after the determination, $\mathbf{\Delta}_{ci}\underline{\mathbf{b}}_{ij}$, namely the confidence interval of the bonding vectors which is found by

$$M(\underline{\underline{ST}}) = \sum_{p=1}^{a} \mathrm{lb}\left(\frac{\mathbf{\Delta}_{ex}\underline{\mathbf{b}}_{ij}}{\mathbf{\Delta}_{ci}\underline{\mathbf{b}}_{ij}}\right)_p \tag{11.12}$$

where $\mathbf{\Delta}_{ex}\underline{\mathbf{b}}_{ij} = \underline{\mathbf{b}}_{ii,\,max} - \underline{\mathbf{b}}_{ij,\,min}$ is the expectation interval, $\mathbf{\Delta}_{ci}\underline{\mathbf{b}}_{ij} = 2s_b t(\alpha, f)/\sqrt{n}$ is the confidence interval of the bonding vectors,[2] and a is the number of the bonds to be considered in the molecule or unit cell. Since, in general, both the expectation and confidence intervals do not differ significantly, Eq. (11.12) is simplified into

$$M(\underline{\underline{ST}}) = a \cdot \mathrm{lb}\frac{\mathbf{\Delta}_{ex}\underline{\mathbf{b}}_{ij}}{\mathbf{\Delta}_{ci}\underline{\mathbf{b}}_{ij}} = \mathrm{lb}\frac{\det(\underline{\mathbf{b}}_{ij})}{\det(\underline{\mathbf{\Delta b}}_{ij})} \tag{11.13}$$

with the matrix of bonding uncertainties $\underline{\mathbf{\Delta b}}_{ij} = (\mathbf{\Delta}_{ci}\underline{\mathbf{b}}_{ij})$.

For finding bond angles that follow directly from a structural systems symmetry or for finding the lattice constants of unit cells,

---

[2] s_b is the total standard deviation of the estimation of the bonding vector from n repeated measurements, and $t(\alpha, f)$ is the quantil of the t-distribution mentioned in Section 6.6.2.

the bonding vectors can be substituted by the bond lengths in Eq. (11.13)

$$M(\underline{\underline{ST}}) = a \cdot 1b\frac{\Delta_{ex}b_{ij}}{\Delta_{ci}b_{ij}} = 1b\frac{\det(b_{ij})}{\det(\Delta b_{ij})} \qquad (11.14)$$

For systems of low symmetry, bond angles are determined in addition to bond distances. Equation (11.13) then becomes

$$M(\underline{\underline{ST}}) = a \cdot 1b\frac{\Delta_{ex}b_{ij}\,\Delta_{ex}\alpha\Delta_{ex}\beta\Delta_{ex}\gamma}{\Delta_{ci}\mathbf{b}_{ij}\,\Delta_{ci}\alpha\,\Delta_{ci}\beta\Delta_{ci}\gamma} \qquad (11.15)$$

Information amounts obtained by Eq. (11.15) are, as a rule, very large. If the expectation intervals $\Delta_{ex}b_{ij} = 0.5$ nm, $\Delta_{ex}a = \Delta_{ex}\beta = \Delta_{ex}\gamma = 180°$, and the confidence intervals $\Delta_{ci}b_{ij} = 0.005$ nm, $\Delta_{ci}a = \Delta_{ci}\beta = \Delta_{ci}\gamma = 1°$, are taken for the estimate, there is an information content of 29.1 bit per bond provided that the kind of bonding is known $[P(b) = 1]$. After its qualitative structure has been determined, a molecule composed of N units may require an information amount as large as

$$M(\underline{\underline{ST}}) = 29.1 \cdot a_N = 15.55 \cdot N(N-1)\, \text{bit} \qquad (11.16)$$

for its quantitative structure analysis. This information amount increases in the case of high-precision determinations of the bond distances and angles.

11.3 INFORMATION AMOUNT OF STRUCTURE-ANALYTICAL METHODS

In practice, we depend above all on spectroscopic methods to provide information about structural groups, the composition of molecules or crystals, and their symmetry. From the point of view of information theory, the analysis of structural groups can be reduced to a problem of elemental analysis, primarily that of identification. Spectroscopic methods provide signals; the signals are characteristic

Table 11.2. Relationships Between Signal Parameters and Analytical Information

Signal Parameter	Elemental Analysis	Analysis of Molecular Structure	Analysis of Crystal Structure
Number	Number of elements of compounds	Molecular symmetry (SR)	Structural type (ER)
Location (energy)	Species (nature of elements or compounds)	Type of structural groups	Type of elementary cell
Intensity	Quantity of the occurring elements or compounds	Quantity of compounds or structural groups (Nature of chemical Bond)	Type of lattice atoms or ions MC: crystal structure PC: texture
Shape	(Multi-component analysis)	Interaction between different structural groups	Real structure, lattice defects

Note: MC: monocrystalline material, SR: selection rules, PC: polycrystalline material, ER: extinction rules.

of the presence of certain functional groups as well as their numbers. Table 11.2 lists some general facts about the signal parameters for elemental or structure analysis.

Information about structural groups is mainly provided by IR–, Raman–, and NMR spectroscopy, from which symmetry relations can also be obtained. The information amount of these and other molecular-spectroscopic methods is described by the equations given in Section 7.3, where in the most general case the potential information amount is

$$M_p = N_p \text{ lb } m \tag{7.24}$$

with the potential analytical resolving power N_p (see Table 7.3):

$$N_p = \int_{z_{min}}^{z_{max}} \Delta z^{-1} \, dz \tag{7.19}$$

A characteristic of constitutional analysis is that the correlation between signals and species is frequently ambiguous. Therefore the different methods of structure elucidation have to be combined. The combination of UV/VIS–, IR–, MNR–, and mass spectroscopy has proved to be especially suited for constitutional analysis because all this information is complementary. From the point of view of information theory, although such combinations of analytical methods may provide redundant information, they improve the reliability of the structural assessments.

For quantitative structural information, the information amounts are generally much larger than for qualitative structural information. Equation (11.16) shows the increase in information required for an increasing number of structural units. For $N = 500$, $M(\underline{\underline{ST}})$ $= 1.8 \cdot 10^6$ bit already exceeds many times the potential information amount of the most effective two-dimensional spectroscopic methods. Such large information amounts are only achieved by analytical methods that provide three-dimensional information of the form $y = f(z_1, z_2)$, as can be seen from Fig. 11.5.

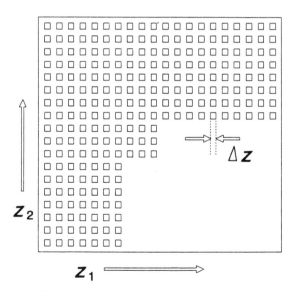

Figure 11.5. Three-dimensional analytical information; the y-axis is perpendicular to the $z_1 z_2$-plane

Here the information amount is derived from

$$M(\underline{\underline{ST}}) = N(\underline{\underline{ST}}) \cdot I_y = N(\underline{\underline{ST}}) \cdot 1b \, m_y \qquad (11.17)$$

where I_y is the information content of an analytical signal. From

$$N(\underline{\underline{ST}}) = N_1 \cdot N_2 = \int_{z_{1,\,min}}^{z_{1,\,max}} \Delta z_1^{-1} \, dz_1 \int_{z_{2,\,min}}^{z_{2,\,max}} \Delta z_2^{-1} \, dz_2 \quad (11.18)$$

we have the two-dimensional analytical resolving power required for quantitative structural analysis. Assuming constant half-peak widths, Eq. (11.18) leads to

$$N(\underline{\underline{ST}}) = \frac{(z_{1,\,max} - z_{1,\,min})(z_{2,\,max} - z_{2,\,min})}{\Delta z_1 \cdot \Delta z_2} \qquad (11.19)$$

In practice, such large information amounts are obtained by diffraction methods. The commonly used procedures of X-ray, electron, and neutron diffraction are based on different interaction principles. While X rays are scattered at the valence electrons of atoms, for which the scattering power strongly depends on atomic number, electrons and neutrons with different cross sections are

Table 11.3. Analytical Resolving Power N_p and Potential Information Amount M_p of Selected Methods of Structural Analysis

Method	N_p	M_p (in bits)
IR spectroscopy	10^3	10^4
Mass spectroscopy	$5 \cdot 10^2$	$5 \cdot 10^3$
• High resolution	$2 \cdot 10^5$	$2 \cdot 10^6$
NMR spectroscopy	10^3	10^4
• High resolution	10^5	10^6
• 2D	10^6	10^7
X-ray diffraction		
• Two-dimensional[a] Debye-Scherrer	$2 \cdot 10^2$	10^3
• Three-dimensional[b] Laue	10^8	10^9
• Precision goniometer	10^{12}	10^{13}
Electron diffraction	10^4	10^5
Neutron diffraction	10^4	10^5

[a]One-dimensional pattern, two-dimensional analytical information.
[b]Two-dimensional pattern, three-dimensional analytical information.

scattered at atomic nuclei. Therefore the information of different diffraction methods is complementary in the same way as qualitative and semiquantitative structural information of molecular-spectroscopic methods. Table 11.3 lists the orders of magnitude of analytical resolving power compared to the information amount of structure-analytical methods.

The information-theoretic fundamentals of structure analysis are pointed out more in detail by Danzer et al. (1976) and Danzer and Marx (1979a, 1979b, 1980). For another perspective, see Bonchev (1985) who applies information theory to structural information.

11.4 PRACTICAL APPLICATIONS

Two classic examples illustrate how quantitative information is obtained by structural analyses of molecules and crystals. For benzene, which from spectroscopic investigations is known to have a planar structure and symmetry c_{6v}, we find the bond lengths to be 0.140 ± 0.001 nm by Eq. (11.14), so the information amount $M(\underline{ST})_{benz} = 6 \cdot lb(0.140/0.001) = 42.8$ bit.

For the lattice constant k_0 of cubic crystals, we get an information amount of $M(\underline{ST}) = a \cdot lb(k_0/\Delta k)$. If $k_0 = 0.200$ nm and $\Delta k = 0.001$ nm are taken as representative average values, we have $M(\underline{ST})_{cub, bc} = 68.8$ bit for body-centered cubic crystals ($a = 9$) and $M(\underline{ST})_{cub, fc} = 107.0$ bit for face-centered cubic crystals ($a = 14$).

REFERENCES

Bonchev, D. 1985. *Information Theoretic Indices for Characterization of Chemical Structure*. Research Studies Press, Letchworth.

Danzer, K., and Marx, G. 1979a. *Chem. analit.* [Warsaw] **24**, 33.

Danzer, K., and Marx, G. 1979b. *Chem. analit.* [Warsaw] **24**, 43.

Danzer, K., and Marx, G. 1980. Proc. Fifth International Symposium on High-Purity Materials in Science and Technology, Dresden. Vol. II: Characterization, p. 166.

Danzer, K., Than, E., Molch, D., and Küchler, L. 1976. *Analytik– Systematischer Überblick*. Akademische Verlagsgesellschaft Geest & Portig, Leipzig; 2d ed. 1988.

CHAPTER

12

CHEMOMETRICS: INFORMATION AMOUNT OF MULTIVARIATE DATA ANALYSIS

Global problems such as preservation of human health, environmental protection, and development of new materials demand very complex analytical systems capable of handling an enormous number of investigations. Because of the increased sophistication of analytical instruments, analysts nowadays are faced with large amounts of data that cannot be evaluated and interpreted exhaustively by conventional ways. For this reason there has developed a separate chemical discipline known as chemometrics whose practitioners employ mathematical and statistical methods to plan, select, execute, and evaluate chemical experiments and analyses in order to extract maximum chemical information from measured data (Kowalski 1980).

The real progress in chemometrical methods, as opposed to classical univariate statistical techniques, is in the use of multivariate statistical methods for the treatment and evaluation of large multidimensional data sets (Deming 1991). The types of investigations in chemometrics are listed in Fig. 12.1. Today chemometrics is a powerful field in analytical chemistry in that it encompasses not only well-known mathematical and statistical techniques such as matrix algebra, numerical methods, linear optimization, and multivariate statistics but also relatively modern techniques such as learning theory, artificial neural networks, genetic algorithms, fuzzy set theory, and the theory of fractals.

Three questions of general interest in multidimensional data sets are answered by chemometric methods:

1. Are there any structures in the data? ⇒ *Cluster analysis* (the search for groups in the data without a priori knowledge about the existence and number of such groups; unsupervised learning techniques)

CHEMOMETRICS			
Experimental design	Data analysis	Correlation analysis	Information theory
Optimization	Factor analysis	Multivariate regression	System theory
Polyopti- mization	Principal component analysis	Soft modeling (PLS)	Signal theory and -processing
Modeling	Pattern recognition	Calibration	Reliability theory
Process control	Classification	Time-series analysis	Operations research
Sampling theory	Cluster analysis		

Figure 12.1. Analytical areas of chemometrics

2. Are the structures—known, expected or found by cluster analysis—reproducible and characterizable quantitatively? ⇒ Pattern recognition[1] (the estimation of a classification model by a training set of data of objects with known categorization in the first step—the learning phase—and classification of unknown objects by means of the found model in the second step—the working phase—; supervised learning techniques)

3. What led to the given data structures? ⇒ *Factor analysis, principal component analysis* (the estimation of factors effective in reality and causing variance in the data; mostly the number of relevant factors or principal components that are smaller than the number of original variables leading to a reduction of the dimensions of the data space)

All three methods enable a *feature reduction* in that previously multidimensional data can be represented in a space of lower

---

[1]The term "pattern recognition" is frequently used as a generic term for both cluster analysis and classification methods. But in a narrow sense only classification methods recognize patterns while cluster analysis cognizes patterns (Derde and Massart 1982).

dimension (frequently in two dimensions). Another method of multivariate data analysis—*PLS modeling and regression* (partial least squares, also projection to latent structures)—introduced in analytical chemistry by Wold and coworkers (Lindberg, Persson, and Wold 1983; Sjöstrom et al. 1983), allows additionally the projection of multidimensional data structures from one data space to another by using latent quantities in reduced dimensions. From the information-theoretic point of view, it is important that in display techniques based on the three methods and others such as nonlinear mapping and the minimal spanning tree, the information from the multidimensional space (m variables) is transformed into a two-dimensional visualization without remarkable loss of information.

Suppose that we have a data matrix of n objects (samples) and m variables (concentrations of m elements). The data are traditionally characterized by m means and m standard deviations, so we have $2m$ parameters. By applying the method of multivariate data analysis, we evaluate the entire variance/covariance matrix. Now we have additionally $m(m - 1)/2$ parameters, namely the covariances. By the covariance terms it is possible to characterize correlations and interactions between the variables. Therefore we obtain the needed information on the internal structure of the data set.

Independent of the aim of a particular method, the information amount of multivariate data analysis relates to that of univariate evaluation by

$$\left[\frac{m(m - 1)}{2} + 2m \right] : 2m \qquad (12.1)$$

Consequently the information amount is higher by a factor $f = (m + 3)/4$ in the multivariate case. The factor by which the information amount increases is as follows: $m = 3 : f = 1.5$; $m = 5 : f = 2.0$; $m = 10 : f = 3.25$; $m = 30 : f = 8.25$; $m = 100 : f = 25.75$. Moreover, if we treat multivariate problems in an univariate way, then the probability P decreases according to

$$P = \prod_{i=1}^{m} P_i \qquad (12.2)$$

where P_i is the chosen probability (significance level) for the problem. Therefore information decreases by more than the factor f when multivariate problems are treated by univariate techniques.

12.1 PATTERN RECOGNITION

When we use *cluster analysis* as an unsupervised pattern recognition technique, we often have no a priori information about any structure of the data set. That means, there is a large number q of possible ways to divide the object into any number of unknown groups, c. Frequently these possibilities have the same probability, $P(c)_0$.

In hierarchical cluster analysis, the points (objects in the m-dimensional feature space) are connected stepwise according to their multidimensional similarities expressed usually by the euclidean distance

$$d_{ij} = \left[\sum_{k=1}^{m} (x_{ik} - x_{jk})^2 \right]^{1/2} \tag{12.3}$$

The hierarchical clustering result is usually given in the form of a dendrogram, a schematic as shown in Fig. 12.2.

From the clustering, there should be observed some structure in the data relevant to the algorithm used (single linkage, average linkage, complete linkage, Ward's method, flexible strategy, etc.; see, e.g., Massart and Kaufman 1983).

The previous uncertainty expressed by the information entropy

$$H_0 = - \sum_{i=1}^{g} \sum_{j=1}^{q} P(c_i)_{0,j} \,\mathrm{lb}\, P(c_i)_{0,j} \tag{12.4}$$

where $g = 1, \ldots, n$ is the number of possible groups (clusters) into which the n objects can be divided and q the number of different ways in which each can be done. If there is no a priori knowledge about the system under investigation, all sizes of clusters are possible, and the n objects may even permute between all groups. The total number of possible classifications, $Z = g \cdot q$, then can be

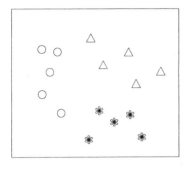

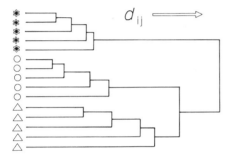

Figure 12.2. The relatedness of 15 objects represented by a dendrogram

calculated by Stirling's second-order recursive formula as applied by Wienke and Danzer (1986).

If we have some a priori information, such as about the number of clusters, then g decreases and consequently H_0 too. In the case of $n = 2$ clusters, for example, the formula of the binomial coefficients yields the same results as Stirling's formula. With knowledge about clustering, the uncertainty decreases and is represented by the a posteriori entropy.

$$H = - \sum_{i=1}^{g} P(c_i) \text{lb} \, P(c_i) \tag{12.5}$$

The probabilities for the c relevant classes found then are much larger than those for the $g - c$ others. On the condition that the a priori probabilities are equal, the information gain I_{PR} of a given classification can be qualitatively characterized (i.e., as corresponding to a certain arrangement) by

$$I_{PR} = -\sum_{i=1}^{c} \frac{n_i}{n} \operatorname{lb} \frac{n_i}{n} = \sum_{i=1}^{c} \frac{n_i}{n} \operatorname{lb} \frac{n}{n_i} \qquad (12.6)$$

where n_i is the number of objects in each c cluster found as the result of cluster analysis (see, e.g., Marengo and Todeschini 1993).

The quality of separations depends on the distances between different clusters in relation either to all the distances or to the distances within the clusters. If we apply the probabilities $P(c)_0$ and $P(c)$ as normalized functions

$$P(c)_s = \frac{g_{s,i} f_{s,i}}{\sum\limits_{i=1}^{Z} g_{s,i} f_{s,i}} \qquad (12.7)$$

where $g_{s,i}$ are arbitrarily chosen weights and $f_{s,i}$ the separation function for the sth state of separation. A derivation of this quantity and its connection to the geometrical parameters of clustering is given by Wienke and Danzer (1986) who use the separation function as a measure of the separation quality of each step and then are able to estimate the relevant number of clusters.

Unlike cluster analysis, in supervised pattern recognition (i.e., classification methods in a narrow sense) the number of classes into which the objects are arranged is known, either by factual reasons or from preliminary information. Frequently, the relevant number of classes is given in a statement of the problem under investigation such as quality in product control (good/sufficient/poor), findings in clinical investigations (healthy/ill; physiological/pathological), provenance of art objects (cultural epoch; genuine/faked), and origin of goods (country; branched/adulterated article). The correct classification is learned by directing sets of data with well-known membership to certain training classes. The different algorithms, like k-nearest neighbor method (kNN), linear learning machine on the basis of linear discriminant functions and Bayesian rules, and SIMCA, are described by Brereton (1990, 1992), Massart et al. (1988), and Sharaf, Illman, and Kowalski (1986).

The efficiency of learning is tested frequently by reclassification of the training set. Among the n objects assigned to c classes (n_i in each, $i = 1, \ldots, c$; $n = \Sigma n_i$), there can occur error-free reclassification or some of the objects may change their class membership. The stability of a classification procedure therefore is expressed by a ratio of correctly to incorrectly classified objects

$$r_c = \frac{n - n_{+/-}}{n} \qquad (12.8)$$

where $n_{+/-}$ is the number of regrouped objects that was originally assigned to a certain class and after reclassification was found in another class. Unlike Eq. (12.6), the real information gain decreases by this factor, so we have

$$I_{PR,r} = r_c \sum_{i=1}^{c} \frac{n_i}{n} \, lb \, \frac{n}{n_i} \qquad (12.9)$$

The classification becomes more reliable, the more distinct the classes are. This fact can be expressed by variance/covariance matrices within the c classes, $\underline{S}_i$, $i = 1, \ldots, c$, and total, $\underline{S}_t$, in the case of supervised pattern recognition, and by adequate distance matrices, $\underline{D}_i$ and $\underline{D}_t$, respectively, in case of cluster analysis. Now the information gain of pattern recognition can be quantitatively characterized as

$$I_{PR,qn} = \sum_{i=1}^{c} lb \, \frac{det(\underline{S}_i)}{det(\underline{S}_t)} \qquad (12.10)$$

in the supervised case. In unsupervised pattern recognition the information gain is adequately

$$I_{PR,qn} = \sum_{i=1}^{c} lb \, \frac{det(\underline{D}_i)}{det(\underline{D}_t)} \qquad (12.11)$$

Minimization of the relations $det(\underline{S}_i)/det(\underline{S}_t)$ and $det(\underline{D}_i)/det(\underline{D}_t)$, respectively, is used in some procedures to find the optimum partition of objects into classes or clusters.

12.2 FACTORIAL DATA ANALYSIS

Factorial data analysis can be carried out either by *factor analysis* (FA) or principal component analysis (PCA). The aim of *factor analysis* is to reproduce the correlations between data into a low-dimensional space so that the inherent structure of the data as well as intrinsic factors that causes this structure are elucidated; *principal component analysis* differs in as much as the variance of the data rather than the correlation between the data is reproduced. An important result of both FA and PCA is how many *significant* factors can be found in the data. Such factors can be interpreted in a chemical or general sense.

Both methods use the decomposition of the original data matrix, $\underline{X}$, into the product of two matrices plus a matrix of residual error:

$$\underline{X} = \underline{T} \cdot \underline{P} + \underline{E} \qquad (12.12)$$

where $\underline{T}$ is the score matrix and $\underline{P}$ the loading matrix of the relevant factors and principal components, respectively. For the so-called reduced factorial solution, the dimension of the latent variable s, extracted by factorial analysis, is low compared with that of the original variable, $s < m$. Accordingly the m-dimensional structure of the data can be represented in a low-dimensional space, frequently in a plane.

The information gain by factorial data analysis can be characterized by

$$I_{FA} = \sum_{i=1}^{s} p_i \operatorname{lb} p_i^{-1} - \sum_{j=1}^{m-s} r_j \operatorname{lb} r_j^{-1} \qquad (12.13)$$

where s is the number of significant factors found, each interpreting a part p_i of the variance of the data set, $(m - s)$ is the number of nonsignificant factors whose variance shares r_j are assigned to the residual error $\underline{E}$. It holds that $\sum p_i + \sum r_j = 1$.

The specific information content of the ith extracted factor is given by

$$I_i = p_i \operatorname{lb} p_i^{-1} \qquad (12.14)$$

In general, it follows that

$$I_1 \geq I_2 \geq \cdots \geq I_s \geq I_{s+1} \geq \cdots \geq I_m \qquad (12.15)$$

so that a graphical representation of p_1 versus p_2 (factors or principal components, respectively) contains a maximum of two-dimensional information about objects, whereas the affiliated scores t_1 versus t_2 comprises information about variables. The fundamentals and methods of factorial data analysis are given, for instance, by Malinowski and Howery (1980), Brereton (1990, 1992), and Sharaf, Illman, and Kowalski (1986). The use of information-theoretic principles in pattern recognition of spectra was shown by Varmuza (1974) and Rotter and Varmuza (1975, 1980) and in the projection pursuit in multivariate analysis by Nason (1992). In general, the information-theoretic fundamentals of chemometric methods can be attributed to the concept of variance reduction of Kateman (1986), see Eqs. (3.38), (3.39).

12.3   PRACTICAL APPLICATIONS

Because of the large size and the high dimension of real data sets, we will look at some simple cases that can be illustrated in two-dimensional representations. Let us consider again Eq. (12.6). Two ways of clustering are now qualitatively compared in Fig. 12.3. We obtain for $n = 35$ case 1 where $n_1 = 17$ and $n_2 = 18$:

$$\begin{aligned} I_{CA} &= \frac{17}{35} \, \text{lb} \frac{35}{17} + \frac{18}{35} \, \text{lb} \frac{35}{18} \\ &= 0.506 + 0.493 \\ &= 1.00 \, \text{bit} \end{aligned}$$

and case 2 where $n_1 = 8$, $n_2 = 7$, $n_3 = 9$, $n_4 = 8$, and $n_5 = 3$:

$$\begin{aligned} I_{CA} &= \frac{8}{35} \, \text{lb} \frac{35}{8} + \frac{7}{35} \, \text{lb} \frac{35}{7} + \frac{9}{35} \, \text{lb} \frac{35}{9} + \frac{8}{35} \, \text{lb} \frac{35}{8} + \frac{3}{35} \, \text{lb} \frac{35}{3} \\ &= 0.487 + 0.464 + 0.504 + 0.487 + 0.304 \\ &= 2.25 \, \text{bit} \end{aligned}$$

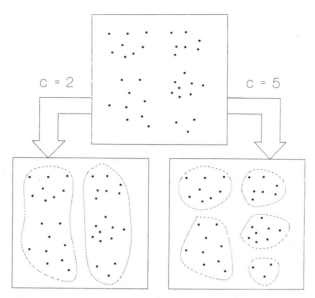

Figure 12.3. Representation of 35 objects in a two-dimensional space of variables with two different ways of clustering

But this information gain is related only to the fact that the data are divided once into two groups and then into five clusters. The goodness of classification is not evaluated in this way.

Goodness of classification can be evaluated by the quantitative measure of information gain, such as by Eq. (12.10). We will demonstrate it for the data presented in Fig. 12.4, using Eq. (12.11). The information gain, characterized as qualitative, is obviously the same in the cases (*a*) and (*b*); namely $I_{CA} = 2[4/8\,\mathrm{lb}(8/4)] = 1.00$ bit.

Quantitatively, with the distance matrices (normalized in both cases to $d_{12} = 1$, where the objects are numbered "line-by-line" from the top left-hand corner to the bottom right-hand corner), we obtain

$$\mathbf{\underline{D}}_1 = \mathbf{\underline{D}}_2 = \begin{pmatrix} 1 & 2 & 3 & 4 \\ \hline 0 & 1.0 & 1.0 & 1.4 \\ 1.0 & 0 & 1.4 & 1.0 \\ 1.0 & 1.4 & 0 & 1.0 \\ 1.4 & 1.0 & 1.0 & 0 \end{pmatrix}$$

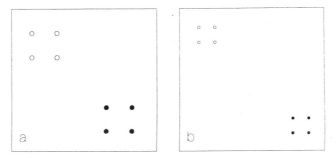

Figure 12.4. Two groups of objects in different separation states

and for case (*a*) of Figure 12.4,

$$
\mathbf{\underline{D}}_{t,a} =
\begin{pmatrix}
1 & 2 & 3 & 4 & 5 & 6 & 7 & 8 \\
\hline
0 & 1.0 & 1.0 & 1.4 & 3.5 & 4.3 & 4.3 & 5.0 \\
1.0 & 0 & 1.4 & 1.0 & 2.9 & 3.5 & 3.8 & 4.3 \\
1.0 & 1.4 & 0 & 1.0 & 2.9 & 3.8 & 3.5 & 4.3 \\
1.4 & 1.0 & 1.0 & 0 & 2.1 & 2.9 & 2.9 & 3.5 \\
3.5 & 2.9 & 2.9 & 2.1 & 0 & 1.0 & 1.0 & 1.4 \\
4.3 & 3.5 & 3.8 & 2.9 & 1.0 & 0 & 1.4 & 1.0 \\
4.3 & 3.8 & 3.5 & 2.9 & 1.0 & 1.4 & 0 & 1.0 \\
5.0 & 4.3 & 4.3 & 3.5 & 1.4 & 1.0 & 1.0 & 0
\end{pmatrix}
$$

and case (*b*),

$$
\mathbf{\underline{D}}_{t,b} =
\begin{pmatrix}
1 & 2 & 3 & 4 & 5 & 6 & 7 & 8 \\
\hline
0 & 1.0 & 1.0 & 1.4 & 7.1 & 7.8 & 7.8 & 8.5 \\
1.0 & 0 & 1.4 & 1.0 & 6.5 & 7.1 & 7.2 & 7.8 \\
1.0 & 1.4 & 0 & 1.0 & 6.5 & 7.2 & 7.1 & 7.8 \\
1.4 & 1.0 & 1.0 & 0 & 5.7 & 6.5 & 6.5 & 7.1 \\
7.1 & 6.5 & 6.5 & 5.7 & 0 & 1.0 & 1.0 & 1.4 \\
7.8 & 7.1 & 7.2 & 6.5 & 1.0 & 0 & 1.4 & 1.0 \\
7.8 & 7.2 & 7.1 & 6.5 & 1.0 & 1.4 & 0 & 1.0 \\
8.5 & 7.8 & 7.8 & 7.1 & 1.4 & 1.0 & 1.0 & 0
\end{pmatrix}
$$

Then for the case (a) we have

$$I_{CA, qn} = lb \frac{\det \underline{\mathbf{D}}_{t,a}}{\det \underline{\mathbf{D}}_1} + lb \frac{\det \underline{\mathbf{D}}_{t,a}}{\det \underline{\mathbf{D}}_2}$$

$$= 2 \, lb \frac{8,358.6}{0.078} = 33.4 \, bit$$

and for the case (b),

$$I_{CA, qn} = lb \frac{\det \underline{\mathbf{D}}_{t,b}}{\det \underline{\mathbf{D}}_1} + lb \frac{\det \underline{\mathbf{D}}_{t,b}}{\det \underline{\mathbf{D}}_2}$$

$$= 2 \, lb \frac{149,334.9}{0.078} = 41.7 \, bit.$$

In agreement with Fig. 12.4 case (b), which is clearer and more reliable, is evaluated by the higher information gain.

As we mentioned earlier, the results of factorial data analysis are characterized by the number of factors or principal components and their variance shares. Let us consider two variants. In the first variant four factors are extracted by FA with the following variance shares: F1: 0.36 (36%), F2: 0.25, F3: 0.15, and F4: 0.12. The information gain according to Eq. (12.13) is

$$I_{FA} = 0.36 \, lb \frac{1}{0.36} + 0.25 \, lb \frac{1}{0.25} + 0.15 \, lb \frac{1}{0.15}$$

$$+ 0.12 \, lb \frac{1}{0.12} - 0.12 \, lb \frac{1}{0.12}$$

$$= 0.531 + 0.500 + 0.411 + 0.367 - 0.367$$

$$= 1.44 \, bit$$

In a second variant, F4 includes an 8% variance. Frequently only factors with variance shares larger than 10% are considered, and attempts are made to interpret them. In this case the information gain will be less than in the first case. We obtain

$$I_{FA} = 0.36\,\text{lb}\,\frac{1}{0.36} + 0.25\,\text{lb}\,\frac{1}{0.25}$$

$$+ 0.15\,\text{lb}\,\frac{1}{0.15} - 0.24\,\text{lb}\,\frac{1}{0.24}$$

$$= 0.531 + 0.500 + 0.411 - 0.494 = 0.95\,\text{bit}$$

An important concern of chemometrics is redundance in large data sets which can diminish the amount of relevant information. For this purpose chemometric methods focus on feature reduction and variance decomposition, and the minimization of variances within classes.

Although chemometric results, particularly from multivariate data analyses, can be evaluated by information theory, as mentioned above, the information gain in this connection is of a different character than that of other analytical methods. In general, the results obtained by chemometrics correspond to the virtual information that is needed to solve problems or to decide alternatives. In this sense chemometric information is condensed information compared with analytical information.

REFERENCES

Brereton, R. G. 1990. *Chemometrics: Applications of Mathematics and Statistics to Laboratory Systems*. Ellis Horwood, Chichester.

Brereton, R. G. (ed.). 1992. *Multivariate Pattern Recognition in Chemometrics, Illustrated by Case Studies*. Data Handling in Science and Technology, vol. 9. Elsevier, Amsterdam.

Deming, S. N. 1991. *Anal. Chim. Acta* **249**, 303.

Derde, M. P., and Massart, D. L. 1982. *Fresenius Z. Anal. Chem.* **313**, 484.

Eckschlager, K., and Stepánek, V. 1979. *Information Theory as Applied to Chemical Analysis*. Wiley, New York, Sec. 6.15.

Kateman, G. 1986. *Anal. Chim. Acta* **191**, 215.

Kowalski, B. R. 1980. *Anal. Chem.* **53**, 112R.

Lindberg, W., Persson, J. A., and Wold, S. 1983. *Anal. Chem.* **55**, 643.

Malinowski, E. R., and Howery, D. G. 1980. *Factor Analysis in Chemistry*. Wiley, New York.

Marengo, E., and Todeschini, R. 1993. *Chemom. Intell. Lab. Syst.* **19**, 43.

Massart, D. L., Vandeginste, B. G. M., Deming, S. N., Michotte, Y., and Kaufman, L. 1988. *Chemometrics: A Textbook*. Elsevier, Amsterdam.

Nason, G. P. 1992. *Anal. Proc.* **29**, 430.

Ritter, G. L., Lowry, S. R., Woodruff, H. B., and Isenhour, T. L. 1976. *Anal. Chem.* **48**, 1027.

Rotter, H., and Varmuza, K. 1975. *Org. Mass Spectrom.* **10**, 874.

Sharaf, M. A., Illman, D. L., and Kowalski, B. R. 1986. *Chemometrics*. Wiley, New York.

Sjöström, M., Wold, S., Lindberg, W., Persson, S., and Martens, H. 1983. *Anal. Chim. Acta* **150**, 61.

Varmuza, K. 1974. *Monatsh. Chem.* **105**, 1.

Varmuza, K. 1980. *Pattern Recognition in Chemistry*. Springer-Verlag, Berlin.

Wienke, D., and Danzer, K. 1986. *Anal. Chim. Acta* **184**, 107.

Wold, S., Albano, C., Dunn, W. J., Edlund, U., Esbenson, K., Geladi, P., Hellberg, S., Johansson, E., Lindberg, W., and Sjöström, M. 1984. *Multivariate Data Analysis in Chemistry*. In Kowalski, B. R., ed., *Chemometrics: Mathematics and Statistics in Chemistry*,

CHAPTER

13

INFORMATION THEORY—THEORETICAL BASIS OF ANALYTICAL CHEMISTRY?

Analytical chemistry differs from other branches of chemistry in both its scope and approach. While the other chemical disciplines are aimed at acquiring knowledge and creating theories in their respective fields, analytical chemistry develops methods and tools necessary to acquire information about the chemical composition, its changes over time, its spatial arrangement, and the structure of molecules and crystals. The methods of analytical chemistry incorporate not only knowledge from chemistry, physics, physical chemistry, and related technical fields but also from mathematics, namely probability theory and mathematical statistics, system theory and information theory, fuzzy set theory, and the theory of measurement. Thus analytical chemistry is more a multidisciplinary field of natural sciences than a pure chemical branch.

Nevertheless, analytical chemistry is an area of chemistry because it deals with the determination of chemical composition. The methods of analytical chemistry allow us to identify chemical compounds, elucidate their structures, and so on. These methods are now indispensable not only in science but also in various human activities of a more practical nature (clinical analysis, environmental protection, materials testing, and so on).

Practical analytical work makes use of diverse chemical reactions and physical interactions. As a consequence practical analytics is markedly differentiated: The various analytical techniques have their own specific physical and chemical underlying principles and corresponding instrumentation. Analysts specialize in particular techniques and organize their conferences and publish their journals around these techniques; they also use somewhat different terminology.

In fact, though constituting a multidisciplinary branch of science and being methodically differentiated, analytical chemistry is mostly regarded as a separate field of science (as is apparent from the majority of contributions sent to a contest in *Fresenius J. Anal. Chem.* in 1992). As in any other independent science, analytical chemistry also has its own theoretical foundations.

As early as a century ago, the theoretical foundations of the analytical chemistry of that time were dealt with by Ostwald (1894). In 1972 Malissa pointed to the information objectives and system basis of analytical methods in relation to their automation. Twenty years later Danzer (1992) demonstrated that the theoretical foundations of analytical chemistry are multidisciplinary. They are, nevertheless, self-consistent because they include a general, unifying "integrating" point of view, namely the commonly recognized fact that the aim of any chemical and physical analysis consists in gaining information. Some contributions to the contest (see *Fresenius J. Anal. Chem.* 1992) even incorporated this aim into a definition and interpretation of analytical chemistry.

General foundations	Specific foundations
CHEMISTRY: * Chemical composition * Structure of molecules	ANALYTICAL METHODS: * based on chemical reactions * based on physical interactions * based on biochemical principles
CHEMOMETRICS: * Probability theory * Mathematical statistics * Regression & correlation * Data analysis * Artificial intelligence	ANALYTICAL OPERATIONS: * Sampling * Separations * Measurement * Calibration, standards * Signal processing
INFORMATION & SYSTEM THEORY	COMPUTER TECHNIQUES: * Hardware * Software
HISTORY & PHILOSOPHY: * Historical development * Paradigma of anal. chem.	LABORATORY MANAGEMENT, AUTOMATION: * LIMS * Robots

Figure 13.1. Theoretical basis of analytical chemistry

Among the multidisciplinary theoretical foundations of analytical chemistry, we can discriminate between the *specific* foundations of individual analytical techniques and the *general* foundations of field as a whole. Both categories are shown in Fig. 13.1.

Of course the physicochemical foundation of an analytical method is important for treating error factors (Eckschlager 1969) in order to facilitate optimization of the analytical procedure and of the associated signal processing (Doerffel, Eckschlager, and Henrion 1989), but the general information theory and system theory foundation influences the choice of analytical method for addressing any—often nonanalytical and even nonchemical—problem. Although the specific fundamentals of chemical and instrumental methods are rather different, some communities can be found between these different fields of analytical chemistry. The analytical process gives a common basis for all analytical branches.

Discipline-specific terminology is an attribute of the exact natural sciences. Quantitation, which is a process of mathematically expressing definitions of terms and notations that had only been qualitatively articulated before, was introduced into analytical chemistry by Kaiser in 1970. Information theory has also proved useful in the description of analytical operations and procedures, as shown by Doerffel (1985). Information theory, combined with the system approach, has enabled an integration of the multidisciplinary foundations of chemical techniques. Both have contributed to the fact that analytical chemistry is now a separate branch of the exact sciences, sometimes designated as *Analytics* or *Analytical Science* in recent time.

REFERENCES

Danzer, K. 1992. *Fresenius J. Anal. Chem.* Mitt.-bl. Fachgr. Anal. Chem. GDCh 4/92, M104.

Doerffel, K. 1988. *Fresenius Z. Anal. Chem.* **330**, 24.

Doerffel, K., Eckschlager, K., and Henrion, G. 1989. *Chemometrische Strategien in der Analytik*. Verlag für Grundstoffindustrie, Leipzig.

Eckschlager, K. 1969. *Errors, Measurements and Results in Chemical Analysis*. Van Nostrand, London.

Kaiser, H. 1970. *Anal. Chem.* **42** (2) 24A; (4) 26A.

Malissa, H. 1972. *Automation in und mit der analytischen Chemie.* Verlag der Wiener Medizinischen Akademie, Vienna.

Ostwald, W. 1894. *Die wissenschaftlichen Grundlagen der analytischen Chemie.* W. Engelmann, Leipzig.

Fresenius J. Anal. Chem. 1992. **343**, 809–835.

INDEX

263